The Mathematics of Juggling

Springer
New York
Berlin
Heidelberg
Hong Kong
London
Milan
Paris
Tokyo

Burkard Polster

The Mathematics of Juggling

With 114 Illustrations

 Springer

Burkard Polster
Department of Mathematics
Monash University
P.O. Box 28M
Victoria 3800
Australia
Burkard.Polster@sci.monash.edu.au

Mathematics Subject Classification (2000): 05-02, 05A05

Library of Congress Cataloging-in-Publication Data
Polster, Burkard.
    The mathematics of juggling / Burkard Polster.
        p. cm.
    Includes bibliographical references and index.
    ISBN 0-387-95513-5 (alk. paper)
        1. Sequences (Mathematics)   2. Juggling—Mathematics.   I. Title
    QA292 .P65 2002
    515′.24—dc21   2002070555

ISBN 0-387-95513-5   Printed on acid-free paper.

Printed in the United States of America.

9 8 7 6 5 4 3 2 1   SPIN 10881157

Typesetting: Pages created by the author using a Springer LaTeX 2e macro package.

www.springer-ny.com

Springer-Verlag   New York Berlin Heidelberg
A member of BertelsmannSpringer Science+Business Media GmbH

To My Jugglable Anu

Preface

Recent history saw an unprecedented rise in amateur juggling. Today, there are hundreds of juggling clubs around the world, there are a number of specialized juggling magazines, and every year thousands of jugglers gather at juggling conventions to practice, exchange ideas, take part in competitions, and have fun together. Juggling is particularly popular among mathematically minded people such as majors in computer science, engineering, mathematics, and physics. In fact, although exact numbers are hard to come by, an estimate that keeps popping up in this context is that up to forty percent of serious jugglers today have a mathematical background; see, for example, [118]. Given this strong interest in juggling in the (hard) scientific community, it is not surprising that many mathematical aspects of juggling have been explored in depth, sophisticated computer programs have been written that simulate juggling, and a number of juggling robots have even been built.

First and foremost, the target audience of *The Mathematics of Juggling* consists of all mathematically minded jugglers. However, mathematics educators, readers of popular mathematical literature, and mathematicians interested in unusual applications of some of their favorite tools and techniques will also be among the readers of this book.

Also included in this book is a chapter dedicated to the mathematics of bell ringing and new connections between bell ringing and toss juggling. Furthermore, this chapter is the most comprehensive introduction to mathematical bell ringing available in the literature and, consequently, many people interested in this ancient art will want to study at least this part of the book.

My aims in writing this book were the following:

- *Serious Mathematics.* Many good books are available that are dedicated to teaching the different juggling tricks and skills; see, for example, [28], [40], and [94]. On the other hand, many popular articles. serious papers, and even Ph.D. theses have been written dealing with different aspects of the mathematics behind these tricks and skills. The main goal of this book is to summarize, connect, and expand the results dealing with mathematical aspects of juggling in the literature.

- *Serious Juggling.* Introduce jugglers to new systematic ways of thinking about juggling and the way they acquire new juggling skills and tricks. First and foremost, this includes a description and analysis of a compact mathematical language for juggling patterns that has already led to the discovery of many new and attractive tricks.

- *Mathematics Education.* Due to the abstract nature of mathematics, mathematicians often have a hard time communicating to the general public that mathematics and, in particular, pure mathematics can be a lot of fun and very useful. Speaking from experience, I know that a talk/performance of mathematical juggling is a perfect ice-breaker in this respect. I hope that this book will lead more scientists to use mathematical juggling in outreach programs and to communicate to the general public that mathematics can be a lot of fun as well as useful in modeling and understanding just about every problem in real life.

- *Turn Jugglers into Mathematicians.* Using juggling as a unifying theme, provide an intellectually stimulating introduction for mathematically wired jugglers to many beautiful results and techniques from a wide range of mathematical disciplines such as combinatorics, graph theory, group theory, knot theory, classical mechanics, and number theory.

- *Turn Mathematicians into Jugglers.* Make people who are able to appreciate and enjoy the mathematics in this book also give the practical side of juggling a go.

- *Bell Ringing.* As with juggling, the art of bell ringing has a strong following in the scientific community, and the appeal of bell ringing and juggling to the scientific mind seems to be very similar. However, not many jugglers seem to know about bell ringing and, vice versa, not many bell ringers also practice juggling. I hope that including and linking the mathematical aspects of these seemingly unrelated pastimes will enrich and complement the overall mathematical menu offered in this book and introduce and awaken the interest of many readers who are only aware of juggling or bell ringing to the respective other activity.

This book is designed to be accessible at different levels of mathematical sophistication. Anybody who is not put off from picking up this book in the first place by the word "mathematics" in the title should be able to get something out of it. At the same time, full proofs, or at least detailed sketches of proofs, have been included for all those people who are able to appreciate "serious" mathematics.

Description of Contents

In the following, I summarize the different chapters and the material covered in this book.

The first chapter consists of a brief introduction to juggling and its history. In particular, it contains a detailed account of the development of mathematical juggling starting around 1985. This also includes brief reviews of a number of freely available computer juggling programs that are based on the mathematical language for juggling patterns developed in this book. These programs illustrate vividly the power of the mathematical approach to juggling by conjuring up virtual jugglers that can juggle any conceivable number of props in infinitely many ways and all this without drops. I highly recommend that you use one or two of these programs side by side with this book. I also take the opportunity to acknowledge the important, yet not very well-known, contributions of many people to the development of mathematical juggling via their articles posted to the newsgroup `rec.juggling`, their development of computer juggling programs, and so forth.

In Chapter 2, we strip away all the flourishes and contortions involved in a periodic juggling pattern in which throws occur at discrete equally spaced moments in time and in which at most one ball is caught and thrown at any given time. What we are left with can be described by a (simple) juggling sequence; that is, a special finite sequence of nonnegative numbers that records the number of beats the individual throws in the pattern stay in the air. We investigate tests based on averages, juggling diagrams, and permutations that allow us to recognize juggling sequences and the number of balls that are used in juggling a given juggling sequence. We prove that any finite sequence of nonnegative numbers, the sum of whose elements is divisible by the length of the sequence, can be rearranged into a juggling sequence. Using juggling cards, we generate all b-ball juggling sequences, derive a formula for the number of b-ball juggling sequences of period p, and describe some close relationships between juggling sequences and elements of certain affine Weyl groups. We introduce state graphs and transition matrices based on which all b-ball juggling sequences of a given maximal throw height can be generated. Furthermore, we derive various results about state graphs such as upper and lower bounds for the lengths of prime loops of maximal lengths, the fact that the states visited by a juggling sequence determine its throws, and the fact that the complement of the b-ball state graph of height h is the $(h - b)$-ball state graph of height h.

In Chapter 3, we consider multiplex juggling sequences, a first generalization of the simple juggling sequences considered in Chapter 2. Unlike in simple juggling sequences, in a multiplex juggling sequence several throws can be made simultaneously on every beat. We find that many of the results that we derived for simple juggling sequences have counterparts for multiplex juggling sequences. Also included in this chapter are several applications that illustrate how to derive, entirely within the framework of multiplex juggling sequences, combinatorial identities involving Gaussian coefficients, Stirling numbers of the second kind, and their q-analogues. Finally, we include a summary of operations that enable us to transform simple and multiplex juggling sequences into other juggling sequences.

All juggling sequences in Chapters 2 and 3 can, at least in theory, be juggled using one hand only. In Chapter 4, we consider multihand juggling in which several multiplex throws can be made simultaneously by a number of different hands. Multihand juggling patterns are described by juggling matrices. Again, we find that virtually all basic results for simple juggling sequences have counterparts for juggling matrices. This includes the different ways to calculate the number of balls required to juggle a juggling matrix, an algebraic test that allows one to check whether or not a given matrix of the appropriate form is a juggling matrix, and results about state graphs of juggling matrices. We also find that most of the usual operations introduced for juggling sequences have counterparts for juggling matrices. However, we also encounter some new operations that allow one to "contract" any juggling matrix into a juggling sequence and to systematically construct all juggling matrices from juggling sequences.

We have a closer look at a number of distinguished subclasses of juggling matrices. This includes the simple juggling matrices in which every hand handles at most one ball at a time and the distributed juggling matrices in which, as the name suggests, on every beat at most one of the hands does something. Cyclic juggling matrices are special distributed juggling matrices. In a cyclic juggling matrix, the different hands take turns throwing the balls in cyclical order. We are particularly interested in cascades and fountains; that is, juggling patterns described by simple cyclic juggling matrices in which all nonzero throws are to the same height and one such throw is made on every beat.

Also included in this chapter are Shannon's fundamental results about uniform multihand juggling patterns. In a uniform juggling pattern, there is not necessarily an underlying beat on which all the throws occur. In this respect, these patterns do not have to be "uniform" at all. What makes them uniform is the fact that the dwell time that any ball is held by a hand is constant, the flight time that any ball spends in the air between being thrown and being caught is constant, and the vacant time that any hand spends empty between throwing and catching is also constant. Shannon's celebrated first theorem relates the dwell, flight, and vacant times of a uniform juggling pattern with the numbers of balls and hands used in juggling the pattern. His second and third theorems deal with the essentially different ways of uniformly juggling a given number of balls with a given number of hands. We state and prove Shannon's theorems and some corollaries and extensions of these results. In particular, we derive generalizations of Shannon's results in the framework of simple cyclic juggling matrices.

Another interesting idea that first pops up in the investigation of uniform juggling patterns is that we never just juggle balls. What we really do is juggle balls and hands. In fact, it follows immediately from the definition of uniform juggling that if we interpret balls as hands and hands as balls, we arrive at a new uniform juggling pattern. This leads to a duality principle for uniform juggling patterns. We also investigate juggling balls and hands in the framework of simple distributed juggling matrices.

In the following section, we show that juggling simple juggling matrices with labeled balls still yields periodic patterns.

In the final section of this chapter, we consider natural decompositions of simple juggling sequences that arise when these juggling sequences are interpreted as simple cyclic juggling matrices. This includes a nice application of the cycle representations of the permutations that are associated with simple juggling sequences.

Chapter 5 is dedicated to practical aspects of mathematical juggling. First, we list and describe a number of juggling sequences that are particularly attractive from a purely visual point of view or useful for teaching purposes. This is followed by a discussion of ways to make juggling easier by taking away gravity in various ways or bouncing balls and descriptions of how these simple ways of juggling have been used in constructing juggling robots. Following this, we model the 2-hand juggling of cascades and fountains to deduce how high and how accurately these patterns have to be juggled. We derive a simple model for club throws, which we then use, in conjunction with previous results, to explain various lining-up effects in juggling 2-hand fountains and cascades. Finally, we summarize the main advantages of the mathematical language for juggling patterns developed in the previous chapters.

Chapter 6 is an introduction to mathematical bell ringing and its connections with juggling. Bell ringers ring the changes. Mathematically speaking, changes are permutations of the bells that are being rung, and ringing the changes is really ringing special sequences of permutations. We call these sequences "ringing sequences." It turns out that ringing sequences can be juggled if we use balls instead of bells and that the simple juggling sequences that correspond to ringing sequences are particularly well-behaved juggling sequences. Furthermore, the sequence of transitions that corresponds to a ringing sequence on b bells can be interpreted in terms of site

swaps and can be used to generate the b-ball simple ground-state juggling
sequences of period b without repetitions. We also give a fairly complete
introduction to mathematical bell ringing. This includes descriptions of
the different operations that enable us to transform ringing sequences into
new ringing sequences, definitions of methods and principles (the two main
basic building blocks for constructing ringable ringing sequences), interpretations of maximal ringing sequences (extents) as Hamiltonian cycles in
Cayley graphs, enumerations of the method- and principle-based extents on
small numbers of bells, a procedure for constructing extents on any number
of bells, and various ways of constructing extents by stringing together left
and right cosets of certain subgroups of symmetric groups.

The final chapter is a collection of articles on juggling- and mathematics-
related topics that do not fit into any of the other chapters. First, we give an
account of recent discoveries of solutions to the 3- and many-body problems.
A number of these solutions are basically cascade juggling patterns. Of
particular interest is a 3-body solution because it is stable and there is
a chance that somewhere in our universe there are really three suns of
approximately equal weight that are being juggled in this way by gravity.

A toss juggler manipulates his props by constantly throwing and catching. In contrast, a contact juggler manipulates his props by rolling them
around his body. In particular, a pyramid of four crystal balls arranged in
the form of a tetrahedron can be spun in one hand, where one of the sides
of the tetrahedron, consisting of three balls, is horizontal to the floor and is
being kept in motion by one hand. The remaining fourth ball rotates much
faster than the three at the bottom. We model this setup and calculate
how fast the top ball can spin.

Many juggling patterns are juggled in a plane in front of the juggler. If
we consider a juggling pattern in three-dimensional plane-time, the set of
trajectories of the balls forms a braid. We show that all finite mathematical
braids can be juggled.

Following this, we show how it is possible to juggle certain words and
rational and irrational numbers, and how the concept of juggling sequences
can be generalized in meaningful ways to juggling sequences that involve
all types of integers, not only nonnegative ones. This also includes a short
introduction to causal diagrams.

The final section of this chapter is a collection of miscellaneous quotes,
anecdotes, puzzles, and so forth that will be of interest to many mathematically inclined jugglers. This includes some information about the most
prominent juggler-mathematician, Ronald Graham.

The appendix lists stereo pictures of Hamiltonian cycles that correspond
to the extents of methods and principles that we considered in the chapter
on bell ringing.

Acknowledgments

This book is based on the work of many mathematicians and jugglers. In particular, I would like to acknowledge the otherwise easily overlooked contributions of all those people who discussed and developed mathematical aspects of juggling on the newsgroup **rec.juggling** as well as all those programmers who spent lots of time and effort turning the ideas that popped up in these discussions into computer juggling programs and made them freely available to the rest of us; see Section 1.3 for more detailed accounts of these contributions.

I am grateful to Kevin Burke for getting me interested in juggling, Bill Baritompa for introducing me to the mathematics of juggling, and Philippe Quoilin for being a constant source of juggling inspiration over the last couple of years.

Particular thanks are due to Theo van Soest for his help with his bell ringing simulator, as well as Arthur T. White and Derryn Griffiths for their help with the mathematical and practical aspects of bell ringing.

I also wish to thank Ina Lindemann at Springer-Verlag for her support of this project, Pam Sayers for her help with the manuscript, Roger Kraft for sending me some very useful articles, and Monash University for its support through a Logan Research Fellowship.

Finally, and most importantly, I would like to like to thank my jugglable Anu for putting up with endless hours of juggling talk and her countless suggestions for improvements of the text and the diagrams.

Melbourne, Australia   Burkard Polster
February 2002

Contents

Preface   **vii**

1   Juggling: An Introduction   **1**
   1.1   What Is Juggling?   1
   1.2   A Very Short History of Juggling   2
   1.3   rec.juggling.   4

2   Simple Juggling   **7**
   2.1   Simplifying Juggling Patterns   7
   2.2   Juggling Diagrams   9
   2.3   Basic Juggling Patterns   11
   2.4   Average Theorem   14
   2.5   Site Swaps and Flattening Algorithm   17
   2.6   Permutation Test   22
      2.6.1   A Method to Construct All Juggling Sequences . . .   24
      2.6.2   Inverse of a Juggling Sequence   25
      2.6.3   Pick a Pattern Procedure   28
      2.6.4   Converse of the Average Theorem   29
      2.6.5   Scramblable Juggling Sequences   34
      2.6.6   Magic Juggling Sequences   35
   2.7   How Many Ways to Juggle?   37
      2.7.1   Juggling Cards   38
      2.7.2   Weights of Juggling Sequences   42
   2.8   Juggling States and State Graphs   44

2.8.1   State Graphs   44
2.8.2   Ground-State and Excited-State Sequences   47
2.8.3   Throws from States   49
2.8.4   Prime Juggling Sequences and Loops   50
2.8.5   Complements of State Graphs   58
2.8.6   Transition Matrices   62

3   Multiplex Juggling   **65**
3.1   Average Theorem and Permutation Test   66
3.2   Number of Multiplex Juggling Sequences   68
3.3   Weights of Multiplex Juggling Sequences   73
3.4   Multiplex State Graphs   75
      3.4.1   Prime Multiplex Juggling Sequences and Loops . . .   77
      3.4.2   Throws from States   81
3.5   Operations Involving Juggling Sequences   81

4   Multihand Juggling   **85**
4.1   Juggling Matrices   85
4.2   Average Theorem and Permutation Test   88
4.3   Multihand State Graphs   90
4.4   Operations Involving Juggling Matrices   92
4.5   Special Classes of Juggling Matrices   94
4.6   Uniform Juggling and Shannon's Theorems   96
4.7   Shannon's Theorems for Juggling Sequences   103
4.8   Cascades and Fountains   107
4.9   Juggling Balls and Hands   110
4.10 Juggling Labeled Balls   112
4.11 Decomposing Simple Juggling Sequences   113

5   Practical Juggling   **117**
5.1   Jugglable Juggling Sequences   117
5.2   Juggling Made Easy   123
      5.2.1   Zero-Gravity Juggling   124
      5.2.2   Bounce Juggling   126
      5.2.3   Robot Juggling   127
5.3   Real-World Juggling with Gravity and Spin   129
      5.3.1   Accuracy and Dwell Time   130
      5.3.2   Why Clubs and Balls Line Up   132
5.4   What Is All this Numbers Juggling Good for?   137

6   Jingling, or Ringing the Changes   **141**
6.1   Enter a Band of Ringers   141
      6.1.1   Basic Definitions   141
      6.1.2   History and Practice of Change Ringing   144
6.2   Juggling the Changes   146

6.2.1   Turning Bells into Balls   146
6.2.2   Turning Extents into Site Swaps   149
6.3   Mathematical Notation and Basic Operations   150
6.3.1   Notation   151
6.3.2   Ringing Sequences from Ringing Sequences   152
6.4   Principles and Methods   154
6.4.1   Principles   154
6.4.2   Methods   155
6.4.3   Extents Based on Principles or Methods   157
6.5   Graphical Representations of Extents   159
6.5.1   Cayley Graphs   159
6.5.2   Four Bells   160
6.5.3   Five Bells   163
6.5.4   Many Bells   166
6.5.5   Names   167
6.6   Extents from Groups   168
6.6.1   Left Cosets and Plain Bob   169
6.6.2   Right Cosets and No-Call Principles   173
6.7   Computers, Bobs, and Singles   175

7   Juggling Loose Ends   177
7.1   Does God Juggle?   177
7.2   Juggling Braids   181
7.3   Spinning Top of a Palm-Spun Pyramid   186
7.4   Useless Juggling   189
7.4.1   Juggling Words   189
7.4.2   Juggling Rational and Irrational Numbers   191
7.4.3   Antiballs, Antithrows, and Causal Diagrams   192
7.5   Juggling and Math Stories   197
7.5.1   Riddle   197
7.5.2   Lord Valentine's Castle   198
7.5.3   Famous Juggler-Mathematicians   199
7.6   Further Reading   199

Appendix: Stereograms of Hamiltonian Cycles   201

References   209

Index   221

1

Juggling: An Introduction

1.1 What Is Juggling?

Most people think of juggling as keeping a number of objects in the air by alternately throwing and catching them. This is also the kind of juggling that we will be focusing on in this book.

However, the word juggling has a number of other meanings as well. For example, you are probably familiar with phrases such as "juggling work and family" or juggling being used in the sense of cheating. Also, rather than trying to define the word "juggling," jugglers will usually describe what they are doing by listing different props, such as balls, clubs, rings, hats, cigar boxes, and diabolos, and the juggling tricks that can be performed with them. There is no complete list, as people are constantly inventing new tricks with old and new props, and there is no widely accepted set of rules among jugglers that defines what counts and what does not count as juggling. Nevertheless, most of the lists used by jugglers to explain juggling will include a vast number of tricks that go beyond the common definition of the word. For example, my personal list also includes tricks such as spinning basketballs on my fingers, balancing a club on my forehead, club swinging, staff twirling, rolling crystal balls around my body, fire breathing, and so forth. However, toss juggling tricks that are usually associated with juggling will occupy a central position on every juggler's list.

In any case, in this book, juggling will mean toss juggling, although other forms of juggling will also be touched upon occasionally.

1.2   A Very Short History of Juggling

Beni Hassan is a burial site in Egypt that includes 150 tombs. Tomb number fifteen is of the Middle Kingdom period of about 1994–1781 B.C. Part of one of the wall paintings in this tomb is the oldest known record of juggling; see Figure 1.1 for a tracing of this part of the painting.

FIGURE 1.1. The earliest known record of juggling is about 4000 years old.

The first juggler is juggling with two balls (her hands are empty), the the second is juggling a 3-ball pattern, and the third juggler is juggling a pattern in which the arms cross. See [45] for a picture of the complete wall painting and more detailed information about this earliest record of juggling and its significance.

Apart from this wall painting, there is plenty of pictorial and written historical evidence that various forms of juggling have been practiced since time immemorial in many parts of the world; see, in particular, Arthur Lewbel's article [78] for a comprehensive survey of the early history of juggling and a good collection of historic depictions of jugglers in action.

For example, the Chinese Book of Lie Zi (475–221 B.C.) mentions a Lan Zi who lived during the Spring and Autumn Period (770–476 B.C.) and was able to juggle seven swords, the Tractate Sukkah of the Talmud reports of a Rabbi Shimon ben Gamaliel who could juggle eight flaming torches, and there is a story of the Irish hero Cuchulainn juggling nine apples (500 A.D.).

The number of objects being juggled in these early accounts is very impressive even by today's standards. Of course, we cannot be sure how accurate these accounts really are and what kind of patterns the early jugglers used. Lewbel points out that it is extremely difficult for someone who is not a juggler to count more than about five objects being juggled. Also, when we talk about juggling, we usually automatically assume that this is done in such a way that every hand throws only one object at a time. However, there is an easier way of getting many objects in the air— by *multiplexing*; that is, having every hand throw several of the objects at the same time. Lewbel also notes that in early depictions of jugglers, the number of objects juggled is usually more conservative; that is, mostly less than six.

Also very interesting are reports from Captain Cook's voyages that talk about Tongan girls juggling up to six balls in a circular pattern, which jugglers refer to as a *shower*. In [112], photographic evidence is cited that seems to prove that there were even girls who were able to juggle ten objects in this way. This feat, if authentic, would beat today's record number of balls juggled in this way, which is eight balls. What makes these accounts more believable is the fact that until recently *hiko*, as juggling is called on Tonga, was practiced by virtually every woman. Juggling four balls was considered standard and five and six balls not uncommon. In such an environment, any juggling talent would be spotted and encouraged at a very early age. For fascinating accounts of juggling on Tonga; see [112], [113], and [114]. We only note that in Japan young girls used to play an old game called *otedama*, which includes singing rhymes and juggling balls, again in a shower pattern.

This is a book about the mathematics of juggling, so we should mention Abu Sahl al-Kuhi, who lived around the tenth century. He started out as a juggler of glass bottles in the marketplace of Baghdad but later gave up juggling to become a famous mathematician; see [12], page 79.

We have very little detailed knowledge about the early jugglers. Towards the end of the Roman empire, juggling was practiced in conjunction with other acrobatic and sleight-of-hand tricks by the traveling *joculatores*. In fact, the word juggling derives from the Latin word *joculare*, which means jesting. Although the joculatores were held in high esteem as artists in the Roman empire, their successors during the Middle Ages were often despised and persecuted just like many other traveling entertainers.

Jugglers were a common sight in marketplaces and the traveling fairs. After the foundation of the first modern circus in 1768, jugglers also became regular acts in these traveling shows. With the advent of vaudeville in the 1880s, jugglers started to appear in theatres and variétés.

Since the early nineteenth century, *strong jugglers* such as Karl Rappo (1800–1854) amazed people by juggling extremely heavy objects such as cannonballs. In the middle of the nineteenth century, some oriental jugglers, such as Mooty and Medua Samme from India, made their forms of elegant juggling popular in Europe. Famous vaudeville jugglers included Paul Cinquevalli (1859–1918), the gentleman juggler Michael Kara (1867–1939), and the comedy juggler W.C. Fields (1880–1946). Enrico Rastelli (1897–1931), who is considered by many the greatest juggler of all time, practiced ten hours a day and was able to juggle ten balls, eight plates, and eight sticks.

With the advent of radio, motion pictures, and television, vaudeville lost more and more of its audience, and by the mid-1950s its demise was complete, many jugglers found themselves out of a job, and street jugglers became a common sight.

Also, soon afterwards, juggling started to become more and more popular as a hobby among young people, and, in particular, among college students. As a consequence, juggling clubs were formed at many colleges and universities, popular compendia of juggling tricks appeared, and a number of national and international juggling organizations were formed. Today, there are tens of thousands of amateur jugglers worldwide who take their hobby very seriously, practice as much or more than the old vaudeville jugglers, and attend juggling conventions in the thousands. The overall shift from professional juggling to amateur juggling has not led to a lowering of standards. In fact, the opposite is true, and the accomplishments of many of today's amateur jugglers surpass those of many vaudeville professionals. We note that club swinging, the precursor to club juggling and passing, had an even larger popular following at the end of the nineteenth century. In fact, it became regular school exercise and even an Olympic discipline for a number of years.

Today, there are still a number of highly successful professional jugglers working in circuses and other high-profile shows. For example, Anthony Gatto, who is considered by many to be the best technical juggler ever, is performing regularly in shows in Las Vegas. He holds many world records and in 2000 was the first juggler to ever win a golden clown at the famous circus festival in Monte Carlo.

For more detailed histories of juggling, see the books and articles [5], [45], [140], [158], and [159] on which this short summary is based.

1.3   rec.juggling

It is not surprising that some of the college students who took up juggling as a hobby also tried to apply the mathematical methods they were studying at the time to describe, analyze, and model the patterns they were juggling in their free time. This led to the invention of a mathematical language of juggling made up of *juggling sequences*, or *site swaps* as they are usually referred to by jugglers, as well as their accompanying juggling diagrams.

It is not completely clear who should be credited with being the first to invent juggling sequences, but it seems that they were discovered independently, packaged slightly differently, and popularized by at least three different groups of people around 1985: Bengt Magnusson and Bruce "Boppo" Tiemann in Los Angeles, Paul Klimak in Santa Cruz, and Adam Chalcraft, Mike Day, and Colin Wright in Cambridge. See also [19], [67], [77], [136], and [153]. for slightly different statements as to who was the first to invent juggling sequences.

In [77], Arthur Lewbel gives a comprehensive survey of the development of juggling notation. He points out, among many other things, that the first published reference to juggling sequences is Magnusson and Tiemann's paper [136], which was published in 1989. However, he also notes that the basic ideas behind juggling sequences were already present in a very short article by Jeff Walker [143], published in 1982, and that Claude Shannon, the famous information theorist, wrote a paper in 1981 on the "Scientific Aspects of Jugging" [118] that contained juggling diagrams of some basic juggling patterns. However, Shannon's paper only got published a decade after it was written. Further references to articles dealing with different kinds of juggling notation include [11], [59], [61], [123], [134], [135], [141], [153], [156], and [157].

Various straightforward extensions of basic juggling sequences have been introduced over the years. One early example is Ed Carstens' *multihand notation* (MHN); see his paper [21] and his computer juggling animator *JugglePro* that is based on this notation.

Many of the questions and results that make up this and the following chapters were first discussed on the newsgroup `rec.juggling`. The most important contributors to this discussion were Jack Boyce, Ed Carstens, Andrew Conway, Dean Hickerson, Allen Knutson, Arthur Lewbel, Bengt Magnusson, Steve Otteson, Willem Rein Oudshoorn, Martin Probert, Steven Rooij, Wolfgang Schebeczek, Jon Stadler, Bruce Tiemann, Johannes Waldmann, and Colin Wright. Most of these jugglers were students in 1985, and without their efforts juggling sequences would not be as well-established a part of juggling and mathematics as they are today. The complete archive of `rec.juggling` can be accessed via the *Juggling Information Service* [62], the most comprehensive collection of juggling resources on the Net. In the following, we will frequently refer to contributions to this newsgroup whenever important questions and results were first mentioned in these articles.

Some of the contributors to the newsgroup were also the first to write computer programs that are able to enumerate "all" juggling sequences and, later on, computer programs that conjure up a virtual juggler whose ability to actually juggle all the corresponding patterns is limited only by the number of virtual balls that fit on a computer screen. The first juggling sequence generator was programmed by Bengt Magnusson in 1991, and the first juggling animator that takes juggling sequences as input was the program *Juggle* by Allen Knutson.

Before you read any further, I recommend that you download (from the *Juggling Information Service* [62]) and play with one of the successors of these programs. Juggling programs are available for free for all platforms. I particularly recommend the program *JuggleAnim* (a JAVA juggling animator) by Jack Boyce. This powerful program is completely platform-independent and all you need to play with it is your (JAVA enabled) Web browser. Its components are JAVA programs that are linked together via a number of Web pages. It contains a generator for juggling sequences, a juggling animator, useful lists of interesting juggling sequences, and good documentation.

I also recommend the following superb juggling animators: *[MA][GNU]S* by Allen Knutson, Greg Warrington, and Matt Levine (JAVA); *JuggleMaster* by Ken Matsuoka (MS-DOS), its MACINTOSH port *MacJuggleMaster* by Chris De Salvo, and its JAVA port *JuggleMaster Java* by Yuji Konishi and Asanuma Nobuhiko; *JoePass!* by Wolfgang Westerboer (MACINTOSH and WINDOWS—download from www.koelnvention.de/jp); and *Jongl* by Werner Riebesel and Martin Hoffmann (with ports by the two authors and Derek Bosch, Pelle Nordqvist, David Byers, and Thomas Ruhroth to pretty much all major platforms except MACINTOSH—download from jongl.home.pages.de). Other good animators with unique features include *JugglePro* by Ed Carstens (MS-DOS and MACINTOSH) and *JuggleKrazy* by Andrew Lipson and Colin Wright (MS-DOS).

2
Simple Juggling

Imagine a juggler. What do you see? A smiling face. A funny hat. Two hands holding an assortment of objects such as balls, knives, and chain saws. Now, imagine a juggler juggling. What do you see? A funny hat wobbling from side to side. All those objects in the air being woven into an array of patterns in the space in front of the juggler. Hands crossing and uncrossing, throwing things behind the back and under the legs. The whole thing finishing with all the objects in the hat, the hat held out towards you, and the smiling face asking the now mesmerized you to add your wallet or at least some of its contents to the objects in the hat. You continue on your way a little poorer, but still enchanted, and trying in vain to figure out exactly what it was that you just witnessed.

A juggler's routine usually consists of a number of tricks or patterns that are strung together for maximum effect. To understand the routine, we have to scrutinize the individual patterns it is composed of, and to understand one of these patterns, we need to break it down into its individual components, such as throws, contortions, and props, and focus on those that are really essential.

2.1 Simplifying Juggling Patterns

Let's ask our juggler to slow down and concentrate on one of the simpler patterns in his routine, skip the contortions, the chain saws, and so forth, and juggle ... very hot potatoes, but let's call them balls anyway. Juggling hot balls discourages the juggler from holding these in his hands for any extended periods of time. This is good because we may then assume that a ball gets caught and thrown at the same point in time. After watching the juggler for a while, we observe the following:

Properties of Simple Juggling Patterns

(J1) The balls are juggled to a constant beat; that is, the throws occur at discrete equally spaced moments in time.

(J2) Patterns are periodic, and we can and will assume that our juggler has been juggling forever and will never stop juggling, repeating the same pattern over and over again.

(J3) At most one ball gets caught and thrown on every beat, and if one is caught, the same ball is thrown.

Conditions J1 and J2 will be satisfied by most juggling patterns considered in this book. Patterns are called *simple* if they also satisfy Condition J3, as opposed to *multiplex patterns* in which hands may catch and throw several balls at the same time. We will consider multiplex patterns in the next chapter. Unless otherwise noted, a juggling pattern in this chapter will be a simple juggling pattern.

When a juggler juggles a juggling pattern, he usually does this by throwing with one of his hands on odd-numbered beats and with the other hand on even-numbered beats. However, at least in theory, every juggling pattern considered in this chapter can also be juggled using just one hand. We "just" have to move this one hand fast enough so that it may play the role of the original right hand on one beat, the role of the original left hand on the next beat, then go back to playing the right hand, and so forth. In practice, even juggling the simplest 3-ball patterns with just one hand is quite hard.

Now, let's have an even closer look at the pattern under consideration. Choose one of the beats to be beat 0, and count the beats up and down from there. We note that the ball that gets thrown on beat

$$\ldots, -4, -3, -2, -1, 0, 1, 2, 3, 4, \ldots$$

stays

$$\ldots, 4, 4, 1, 4, 4, 1, 4, 4, 1, \ldots$$

beats in the air, respectively, before it gets caught and thrown again. As indicated, this pattern is periodic. This means that any of the finite sequences

$$441, 414, 441441, 144144144, \ldots$$

captures the essential information about the pattern; just repeat any of these finite sequences infinitely often forwards and backwards in time to reconstruct the doubly infinite sequence above. We call a finite sequence of nonzero integers arising from a pattern in this way a *juggling sequence.*

We call a throw that lasts h beats an *h-throw* or a *throw of height h.* To execute a 0-throw on a beat just means that no ball gets caught or thrown on that beat. A 1-throw is a straight, more or less horizontal throw from one hand to the other. In theory, a 2-throw is a very low throw from one hand that is caught by the same hand. However, in practice, jugglers perform a 2-throw by just holding a ball for two beats in one hand while the other hand automatically deals with any possible incoming ball.

Examples

Unless otherwise highlighted, you may always assume that in any of our examples throws are at most of height 9. This means that a juggling sequence in an example such as 441 is an abbreviation of 4,4,1 rather than, for example, the juggling sequence 44,1 consisting of throws of height 44 and 1.

To emphasize that we are dealing with something juggling-related, we call the length of a finite sequence of integers its *period.* For example, 441 has period 3. A juggling sequence is *minimal* if it has minimal period among all the juggling sequences representing the same pattern. This means that the sequence 144 is minimal, but 441441 is not. Clearly, the period of any juggling sequence corresponding to one pattern is a multiple of the period of one of its minimal juggling sequences. Furthermore, all juggling sequences of the same pattern of a given period p are cyclic permutations of each other and, consequently, there are at most p such juggling sequences. For example, our sample pattern has the three minimal juggling sequences 441, 414, and 144.

2.2   Juggling Diagrams

In a *juggling diagram* of a juggling pattern or sequence, the displacements of the balls in the vertical direction are plotted with respect to time. Figure 2.1 is the juggling diagram of our sample sequence. Odd- and even-numbered beats are marked by solid and open circles, respectively, which in turn correspond to the left and right hands catching and throwing. Note that in an odd-numbered throw, a ball is thrown from one hand to the other hand, whereas even-numbered throws are thrown and caught by the same

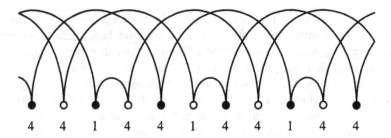

FIGURE 2.1. Juggling diagram of 441.

hand. The larger the number of a throw, the higher and wider the arc that corresponds to it. Note that juggling diagrams are usually not drawn to scale as they, like the juggling sequences they represent, are only used to capture the essence of a pattern.

By the way, how many balls does our juggler use to juggle 441? To answer this question, just note that in the diagram, a ball traces a continuous orbit. Since the diagram is made up of three such orbits, we deduce that three balls are used to juggle the pattern. Alternatively, consider a vertical line that does not contain any of the points of intersection of two of the arcs in the juggling diagram. Then, the number of points of intersection of this line with the arcs in the diagram clearly equals the number of balls juggled. We have also marked below every circle/beat in the diagram what kind of throw is performed on that beat.

One of the easiest ways of checking whether a finite sequence of nonnegative numbers is a juggling sequence is to draw the corresponding would-be juggling diagram. For example, let's check that a sequence of the form

$$n(n-1)\cdots$$

cannot be a juggling sequence. Start by drawing the circles that mark the beats, draw in the first arc spanning n intervals, and draw a second arc spanning $n-1$ intervals. At this point, you will notice that the corresponding two balls have to be caught by the same hand, which is a violation of condition J3; see Figure 2.2 for the resulting collision in the case $n = 3$.

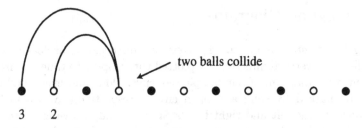

FIGURE 2.2. A sequence of the form $n(n-1)\cdots$ is not a juggling sequence.

Also, the only way a would-be juggling diagram constructed from a sequence as above may not work out is by violating condition J3; that is, we only have to check that on every beat either no ball is being caught and thrown or exactly one is caught and immediately thrown again. Obviously, we do not want to check this for every single beat. So, how many are sufficient? This depends very much on the numbers in the sequence and its period. We can only check whether a 100-throw gets into trouble by at least checking 100 beats past the spot where this throw is performed the first time. More precisely, starting with some beat, we have to check that things pan out in a part of the would-be juggling diagram that fully contains all the arcs corresponding to all throws performed in the period following our starting beat. If p or h is large, then this is not a very practical approach. However, if you are a juggler who quickly wants to check out whether or not to try an interesting-looking sequence (interesting = relatively short and consisting of throws that are not too high), then drawing the juggling diagram is a perfectly feasible thing to do. We will deduce a very quick algebraic test for juggling sequences on page 22.

2.3 Basic Juggling Patterns

To be honest, patterns corresponding to the juggling sequence 441 are not the kind of juggling patterns that the ordinary juggler would be familiar with. However, it is one of the simplest (interesting) juggling patterns that were discovered after the introduction of juggling sequence notation and its subsequent application to enumerate "all" juggling patterns.

The most basic b-ball pattern corresponds to the equally basic single-element juggling sequence b. Depending on whether b is odd or even, this pattern is called the *b-ball cascade* or the *b-ball fountain*. This distinction comes from the already mentioned fact that odd-numbered throws make balls go from one hand to the other while even-numbered throws force the balls to get thrown and caught by the same hand. This means that a fountain really splits into two completely independent parts: exactly $b/2$ balls being juggled by the right hand and the same number of balls being juggled by the left hand. Figure 2.3 shows the juggling diagram of the 3-ball cascade.

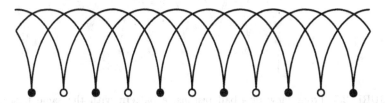

FIGURE 2.3. Juggling diagram of the 3-ball cascade.

In a cascade pattern, all balls travel on a path that is reminiscent of an infinity sign. Figure 2.4 shows a snapshot of a 3-ball cascade and a 4-ball fountain plus the paths on which the balls travel.

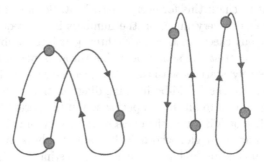

FIGURE 2.4. Front views of the 3-ball cascade and the 4-ball fountain.

Note that these pictures are pretty close to what you actually would see when you watch a real juggler perform these tricks. In particular, jugglers usually do not juggle in hot-potato style. Rather, there is a pretty pronounced carry phase between the catch of a ball and its release. Except for making juggling diagrams more complicated, taking such a carry phase into account would not change anything in our analysis.

We should also stress again that any juggling sequence corresponds to many different patterns that involve various contortions of the body such as varying hand positions. For example, the juggling sequence *b* also corresponds to the pattern *b-ball pistons*. In this pattern, the balls are assigned equally spaced straight vertical paths in a plane in front of the juggler along which they move up and down. Although this juggling pattern looks quite different from a cascade or fountain pattern, its minimal juggling sequence is still *b*. Figure 2.5 shows a snapshot view of 4-ball pistons together with the juggling diagram of the juggling sequence 4.

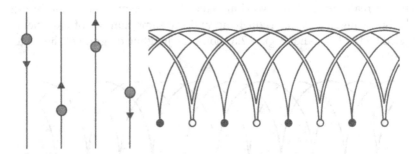

FIGURE 2.5. Front view of 4-ball pistons, a pattern with the same juggling diagram as the 4-ball fountain. Balls never cross between the two hands.

At first glance, pistons may appear to be much simpler patterns than the cascade and fountain patterns. They are not. When we switch from juggling a 4-ball cascade to 4-ball pistons, what we are really doing is switching from a pattern in which the hands basically stay stationary and the balls are being juggled to a pattern in which the balls are fixed in a certain sense and the hands are being juggled. It turns out that in practice the latter is usually a little bit more difficult. In particular, when juggling pistons with an odd number of balls, the two hands have to continuously cross and uncross, which is not easy at all.

Similar observations can be made for all other patterns that have b as their juggling sequence. Even reversing the directions in which the balls traverse a basic cascade or fountain pattern results in patterns that are much harder to juggle than the cascade or fountain. Note that in both cascade and fountain patterns, balls are thrown from the inside to the outside of the patterns. In a *reverse cascade* or *reverse fountain* pattern, balls are thrown from the outside to the inside of the pattern, which, when first attempted, results in a lot of collisions in the middle of the pattern.

In general, given any juggling sequence, there is the most basic pattern that realizes it. In it, the hands are held as in the case of the cascade or fountain patterns, and any b-throw is made roughly as the respective throw in the b-ball fountain or cascade. All other patterns that correspond to the same juggling sequence are usually more difficult to juggle than this basic pattern. The reasons for this are basically the same as those given above in the special case of the one-element juggling sequences.

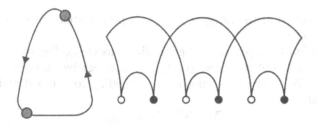

FIGURE 2.6. Front view and juggling diagram of the 2-ball shower.

To conclude this section, let us mention the patterns that most non-jugglers consider as the prototype juggling patterns. The *b-ball shower* has minimal juggling sequence $(2b-1)1$. In this pattern, one hand only performs $(2b-1)$-throws while the other hand performs only 1-throws. Balls travel along a "circular" path; see Figure 2.6 for pictures of the 2-ball shower.

The 2-ball shower is what many people seem to be able to do: Throw one ball in the air with your right hand and zap the other ball from left to right and repeat ad infinitum. It is also this pattern that gets in the way of learning the fundamental 3-ball cascade, which is started by throwing the first ball exactly like the first ball in a 2-ball shower. For people who are very good at 2-ball showers, things then tend to go wrong from there.

2.4   Average Theorem

How many balls are necessary to juggle a given juggling sequence? We have already given one answer to this question: The number of balls equals the number of orbits in the corresponding juggling diagram. Can this number be infinite? On closer inspection of our rules, you will notice that this possibility has not been excluded. Let's discuss this question in a slightly more general setting than considered thus far.

In the following, $\mathbf{Z}$, $\mathbf{N}$, and $\mathbf{N}^0$ will denote the set of all integers, the set of all positive integers, and the set of all nonnegative integers, respectively.

Let's start with a function $j : \mathbf{Z} \to \mathbf{N}^0$. If we consider $\mathbf{Z}$ as the set of beats that underlies our juggling efforts and juggle a $j(i)$-throw on beat i for all $i \in \mathbf{Z}$, then the infinite sequence

$$\cdots j(-3)j(-2)j(-1)j(0)j(1)j(2)j(3) \cdots$$

can be juggled such that Condition J3 is satisfied if and only if the corresponding juggling diagram (defined exactly as in the case of juggling sequences) works out. It is easy to check that this is the case if and only if the function

$$j_+ : \mathbf{Z} \to \mathbf{Z} : i \mapsto i + j(i)$$

is a permutation of the integers. If j_+ is a permutation, then we call j a *juggling function*.

Juggling functions are natural generalizations of juggling sequences. Indeed, given a finite sequence $\{a_k\}_{k=0}^{p-1}$ of nonnegative integers—that is, numbers in $\mathbf{N}^0$—then this sequence is a juggling sequence if and only if the function

$$\mathbf{Z} \to \mathbf{N}^0 : i \to a_{i \bmod p}$$

is a juggling function. For example, the juggling function that corresponds to the juggling sequence b is the constant function $\mathbf{Z} \to \mathbf{N}^0 : i \to b$, and the juggling function that corresponds to the juggling sequence 51 takes on the values 5 and 1 for all even and odd i, respectively. There are juggling functions that do not arise from juggling sequences in this manner. Such a function is

$$\mathbf{Z} \to \mathbf{N}^0 : i \to \begin{cases} 0 & \text{if } i = 0, \\ 2^{T(i)+1} & \text{if } i \neq 0, \end{cases}$$

where $T(i)$ is the highest power of 2 that divides i; see [19]. Figure 2.7 shows the corresponding juggling diagram. Note that, unlike juggling diagrams arising from juggling sequences, this diagram is not periodic.

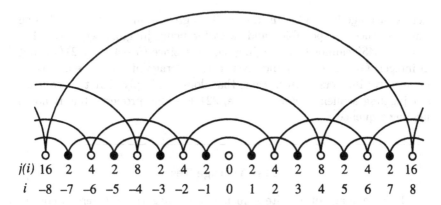

$j(i)$	16	2	4	2	8	2	4	2	0	2	4	2	8	2	4	2	16
i	−8	−7	−6	−5	−4	−3	−2	−1	0	1	2	3	4	5	6	7	8

FIGURE 2.7. A juggling diagram that does not correspond to a juggling sequence. It consists of infinitely many orbits. The heights of the throws have no upper bound.

If j is a juggling function, we define its *height* and give the number of orbits in the corresponding juggling diagram a name:

$$height(j) \;=\; \begin{cases} \infty & \text{if } j \text{ is not bounded,} \\ \max\{j(i) \mid i \in \mathbf{Z}\} & \text{otherwise,} \end{cases}$$

$$balls(j) \;=\; \text{number of orbits of the juggling diagram of } j.$$

Clearly, if j arises from a juggling sequence as above, then the height of j is the maximum height of a throw in the juggling sequence. If j is our last example, you can easily convince yourself that $height(j) = balls(j) = \infty$.

The Average Theorem

(B1) The number of balls necessary to juggle a juggling sequence equals its *average*.

(B2) Let j be a juggling function. If $height(j)$ is finite, then

$$\lim_{|I| \to \infty} \frac{\sum_{i \in I} j(i)}{|I|}$$

exists, is finite, and is equal to $balls(j)$, where the limit is over all integer intervals $I = \{a, a+1, a+2, \ldots, b\} \subset \mathbf{Z}$, and $|I| = b - a + 1$ is the number of integers in I.

Clearly, Result B1 is an immediate consequence of B2. Note that B1 implies that if a finite sequence of nonnegative integers is a juggling sequence, then its average has to be an integer. This gives a quick way of excluding some sequences as possible candidates for being juggling sequences. For example, 5432 cannot be a juggling sequence since $(5+4+3+2)/4$ is not an integer. On the other hand, even if the average of a finite sequence of nonnegative integers is an integer, that does not imply that the sequence is a juggling sequence. For example, 321 has this property but is not a juggling sequence.

The Average Test

If the average of a finite sequence of nonnegative integers is not an integer, then the sequence is not a juggling sequence.

Before we prove B2, let us remark that if $height(j)$ of a juggling function j is infinite, then $balls(j)$ can be infinite, as in the example above. However, this number may also be finite. Consider the juggling function j that corresponds to the juggling diagram in Figure 2.8.

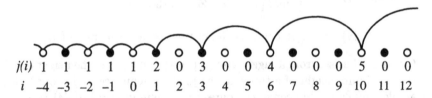

$j(i)$	1	1	1	1	1	2	0	3	0	0	4	0	0	0	5	0	0
i	−4	−3	−2	−1	0	1	2	3	4	5	6	7	8	9	10	11	12

FIGURE 2.8. A juggling diagram that does not correspond to a juggling sequence. It consists of only one orbit. The heights of the throws have no upper bound.

Here, $j(i) = 1$ for all nonpositive beats. We define

$$j\left(\sum_{k=1}^{n} k\right) = n+1$$

for all $n \in \mathbf{N}$. For a positive beat i for which $j(i)$ is still undefined, we set $j(i) = 0$. Clearly, the height of the corresponding juggling function is infinite. However, only one ball gets juggled. Note also that the limit in B2 does not exist for this function, as there are arbitrarily large intervals of negative integers at which the function takes on the constant value 1, as well as arbitrarily large intervals of positive integers at which the function takes on the constant value 0.

Proof of B2. Consider an interval I such that $|I| > height(j)$. Then, for every orbit of the juggling diagram, there is at least one beat in I that contributes to this orbit. This means that $balls(j)$, the number of orbits, is finite. The sum of the values $j(i)$ corresponding to the beats $i \in I$ that contribute to a fixed orbit of the juggling diagram is bounded from below by $|I| + 1 - height(j)$ and from above by $|I| - 1 + height(j)$; see Figure 2.9 for the picture to keep in mind.

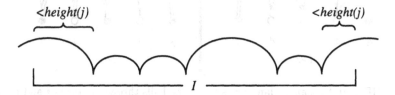

FIGURE 2.9. Upper and lower bounds.

This implies that

$$\frac{balls(j)(|I| + 1 - height(j))}{|I|} \leq \frac{\sum_{i \in I} j(i)}{|I|} \leq \frac{balls(j)(|I| - 1 + height(j))}{|I|}.$$

The expressions on both sides of this chain of inequalities tend to $balls(j)$ as $|I|$ tends to infinity. This proves B2.   □

This is the proof given in [19], Theorem 1. In Subsection 2.6.4, we will prove a partial converse of the Average Theorem.

2.5 Site Swaps and Flattening Algorithm

In this section, we introduce two operations that transform juggling sequences into new juggling sequences and leave important properties of such sequences unchanged. We then use these operations to generate all juggling sequences from the basic ones.

Suppose you are juggling along according to a juggling function j. On beat i, you toss a ball. This means that $j(i)$ is at least 1. Pick a later beat $i+d$ at which or after which the ball lands; that is, $d \leq j(i)$. Then, we can construct a new juggling function j' that coincides with j on all beats except i and $i+d$. The balls thrown according to j' on beats i and $i+d$ will land on beats $j(i+d)$ and $j(i)$, respectively. Put more concisely, the "sites" at which these balls land "swap" positions; see Figures 2.10 and 2.11 for the four essentially different start and end positions for such a transformation. Note that if we apply the same site-swapping operation to j', we end up reconstructing j. This means that all four diagrams in the two figures can serve as starting positions for the site-swapping operation. Explicitly, we find that $j'(i) = j(i + d) + d$ and $j'(i + d) = j(i) - d$. Note that we have made sure that the last number is always nonnegative.

Next, we observe that $balls(j) = balls(j')$. To see this, just note again that the four diagrams in the figures depict the essentially different cases

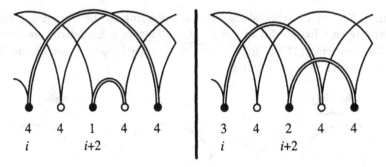

FIGURE 2.10. Swapping (landing) sites of the balls thrown on beats i and $i+2$ before and after the swap.

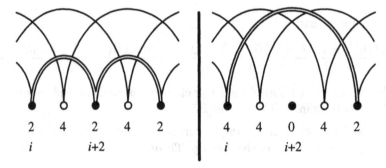

FIGURE 2.11. Swapping (landing) sites that both belong to the same orbit.

that have to be considered. Furthermore, the two arcs that get modified in either of the diagrams in Figure 2.10 have to belong to two different orbits, so, clearly, the number of balls juggled does not change in these cases. Similarly, in both diagrams of Figure 2.11, we are only modifying the shape of one orbit.

Now, we want to apply a similar site-swapping transformation to juggling sequences. Consider the juggling sequence

$$a_0 a_1 \cdots a_{i-1} \boxed{a_i} a_{i+1} \cdots a_{i+d-1} \boxed{a_{i+d}} a_{i+d+1} \cdots a_{p-1}.$$

If $0 < d \le a_i$, then a site swap can be applied to the juggling function associated with this sequence at any of the infinitely many pairs of beats that correspond to a_i and a_{i+d}. Of course, applying the site swap to only one of the pairs will result in a function that is no longer periodic; that is, no longer a juggling function of a juggling sequence. However, since successive site swaps at the different pairs will always give new juggling functions, we may apply site swaps simultaneously at all pairs to arrive at a juggling function of a new site swap. The corresponding new juggling sequence is

$$a_0 a_1 \cdots a_{i-1} \boxed{a_{i+d} + d} a_{i+1} \cdots a_{i+d-1} \boxed{a_i - d} a_{i+d+1} \cdots a_{p-1}.$$

So, nothing changes except for the entries in the boxes. Here is a summary of the most important properties of this operation.

Site Swaps

Let $s = \{a_k\}_{k=0}^{p-1}$ be a sequence of $p \geq 2$ nonnegative integers. Let i and j be nonnegative integers such that $0 \leq i < j \leq p - 1$ and $j - i \leq a_i$. Then, $s_{i,j}$ is the sequence of p nonnegative integers that coincides with s except that the ith and jth elements in the sequence are $a_j + j - i$ and $a_i - j + i$, respectively. We call the operation of transforming s into $s_{i,j}$ a *site swap*[1] of beats i and j of s. Then, the following hold:

(S1) The sequence s is a juggling sequence if and only if $s_{i,j}$ is.

(S2) The average of s is the same as the average of $s_{i,j}$.

(S3) If s is a juggling sequence, then the number of balls used to juggle it is the same as the number of balls used to juggle $s_{i,j}$.

As an example, consider the juggling sequence 642. By performing a site swap of beats 0 and 1, we construct the juggling sequence 552 and, by performing a site swap of beats 0 and 2, we arrive at the juggling sequence 444.

We check that Property S1 is satisfied. First, note that if we perform a site swap of beats i and j of $s_{i,j}$, then we end up with the sequence s we started with. Furthermore, if s is a juggling sequence, then we can check the effect of a site swap on s by inspecting how the bottom part of the juggling diagram of s is affected by the site swap (using Figures 2.10 and 2.11). It follows that $s_{i,j}$ is also a juggling sequence. This shows that s is a juggling sequence if and only if $s_{i,j}$ is. Property S2 follows from the definition of site swaps. Property S3 follows immediately from Result B1 (the number of balls in a sequence equals its average).

Cyclic Shifts

Let $s = \{a_k\}_{k=0}^{p-1}$ be a sequence of $p \geq 2$ nonnegative integers and let $s_\rightarrow$ be the sequence $a_{p-1} a_1 a_2 \cdots a_{p-2}$. We call the operation of transforming s into $s_\rightarrow$ a *cyclic shift* of s. Then, C1, C2, and C3, the counterparts to S1, S2, and S3, hold (replace $s_{i,j}$ by $s_\rightarrow$).

---

[1]Note that, unlike us, jugglers usually refer to juggling sequences as *site swaps*.

We proceed to describe an algorithm invented by Allen Knutson that "flattens" finite sequences of nonnegative integers; see [67].

This *flattening algorithm* takes as input an arbitrary sequence $s = \{a_k\}_{k=0}^{p-1}$ of $p \geq 1$ nonnegative integers. It then uses the following program to transform s into a new sequence of period p:

1. If s is a constant sequence, stop and output this sequence. Otherwise,

2. use cyclic shifts to arrange the elements of s such that one of maximum height, say e, comes to rest at beat 0 and one not of maximum height, say f, comes to rest at beat 1. If e and f differ by only 1, stop and output this new sequence. Otherwise,

3. perform a site swap of beats 0 and 1. Redefine s to be the resulting sequence, and return to Step 1.

Whenever the program performs a site swap, it lowers the number of throws of maximum height in the current sequence. This implies that it will definitely terminate after a finite number of steps. Because of S1 and C1, all intermediate sequences created by the program are or are not juggling sequences if and only if the input sequence is or is not a juggling sequence, respectively. This implies that on input of a juggling sequence, the program will never terminate at Step 2 because e and f will always differ by at least 2; otherwise, we would get a collision as in Figure 2.2. If the average of the input juggling sequence is b and its period is p then Conditions S2 and C2 guarantee that the output sequence will be the *b-sequence of period p*; that is, the sequence of period p all of whose elements are equal to b. On the other hand, if the input sequence is not a juggling sequence, the program has to terminate at Step 2 and the output sequence is of the form $e(e-1)\cdots$.

The following example illustrates how the flattening algorithm transforms the juggling sequence 642 into a constant sequence. Note, again, that all intermediate sequences are also juggling sequences.

$$642 \xrightarrow{\text{swap}} 552 \xrightarrow{\text{2shift}} 525 \xrightarrow{\text{swap}} 345 \xrightarrow{\text{shift}} 534 \xrightarrow{\text{swap}} 444.$$

Here is an example that illustrates what happens to the nonjuggling sequence 514 when we try to flatten it. Note that 514 does not even pass the average test and hence is definitely not a juggling sequence.

$$514 \xrightarrow{\text{swap}} 244 \xrightarrow{\text{shift}} 424 \xrightarrow{\text{swap}} 334 \xrightarrow{\text{shift}} 433 \text{ collision!}$$

Since every step in the flattening algorithm can be reversed, using only site swaps and cyclic shifts, we also arrive at the following nice result:

**Site Swaps and Cyclic Shifts Generate
Juggling Sequences**

The b-sequence of period p can be transformed into any b-ball
juggling sequence of period p by using only site swaps and cyclic
shifts.

There are 37 3-ball juggling sequences of period 3. Up to cyclic shifts,
there are 13. Figure 2.12 shows a *site swap graph* of these juggling sequences. In this graph, juggling sequences that only differ by cyclic shifts
are combined into one vertex in the obvious manner. To translate one of
these vertices into one of the juggling sequences it represents, start at one
of the numbers in the vertex and then read in the counterclockwise direction. For example, the vertex in the lower-left corner represents the three
sequences 900, 090, and 009. The graph has a total of 13 vertices. Two
vertices are connected if the corresponding juggling sequences are one site
swap of two adjacent beats apart. This means that the flattening algorithm
applied to any of the sequences in this graph moves along the edges of the
graph.

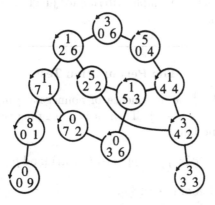

FIGURE 2.12. The 13 (up to cyclic shifts) 3-ball juggling sequences of period 3.

In general, we can draw a similar site swap graph that has as its vertices,
and relates, the b-ball juggling sequences of period p. For $p = 1$, this graph
consists of only one vertex. For $p = 2$, it is a string as in the specific example
depicted in Figure 2.13.

For all choices of $p \geq 2$, these graphs have two "ends" each; that is,
two vertices that are tied into the graph by only one edge. These vertices
are represented by the constant b-sequence of period p and the juggling
sequence $(pb)00 \cdots 0$. It takes the flattening algorithm exactly $b(p-1)$ site

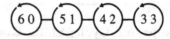

FIGURE 2.13. The four (up to cyclic shifts) 3-ball juggling sequences of period 2.

swaps to flatten this last sequence. Every other b-ball juggling sequence of period p is flattened using fewer site swaps.

Finally, note that the flattening algorithm gives a slick proof that the average of a juggling sequence is the number of balls juggled; see also [19] and [67]. Just observe that this is definitely the case for constant sequences and that site swaps and cyclic shifts leave the number of balls juggled and the average invariant.

2.6   Permutation Test

We already noted that a finite sequence of nonnegative integers is a juggling sequence if and only if a certain associated function $\mathbf{Z} \to \mathbf{Z}$ is a permutation; see page 14. To actually check whether or not we are dealing with such a permutation is equivalent to verifying that the juggling diagram of the sequence pans out. We already remarked that for long sequences or sequences containing large numbers, this is not a very practical approach. In this section, we derive a simple criterion for juggling sequences based on finite permutations.

The Permutation Test

Let $s = \{a_k\}_{k=0}^{p-1}$ be a sequence of nonnegative integers and let $[p] = \{0, 1, 2, \ldots, p-1\}$. Then, s is a juggling sequence if and only if the function

$$\phi_s : [p] \to [p] : i \mapsto (i + a_i) \bmod p$$

is a permutation of the set $[p]$.

The function ϕ_s is a permutation if and only if the *test vector*

$$(\phi_s(0), \phi_s(1), \phi_s(2), \ldots, \phi_s(p-1))$$

contains every one of the integers from 0 to $p-1$.

As a first example, let's consider the juggling sequence 6424. Here, $p = 4$, and

$$(0 + 6, 1 + 4, 2 + 2, 3 + 4) \bmod 4 = (6, 5, 4, 7) \bmod 4 = (2, 1, 0, 3)$$

indeed contains every integer from 0 to 3. We already know that 514 is not a juggling sequence. Indeed,

$$(0 + 5, 1 + 1, 2 + 4) \bmod 3 = (5, 2, 6) \bmod 3 = (2, 2, 1)$$

does not contain all the integers from 0 to 2.

The proof of the result above is again based on the flattening algorithm that we introduced in the last section.

Proof. Let $s = \{a_k\}_{k=0}^{p-1}$ be a sequence of $p \geq 1$ nonnegative integers. In the case $p = 1$, there is nothing to show. Let $p \geq 2$.

We first consider the effect of a site swap of beats i and $i + d$ on the test vector of s. We note that except for the ith and $(i + d)$th entry, no entries of the test vector change. The ith entry $i + a_i \bmod p$ and $(i + d)$th entry $i + d + a_{i+d} \bmod p$ change to $i + (a_{i+d} + d) \bmod p = i + d + a_{i+d} \bmod p$ and $(i + d) + (a_i - d) \bmod p = i + a_i \bmod p$, respectively. This means that the two entries just swap places and the transformed test vector contains all the elements of $[p]$ if and only if the original test vector does.

Second, consider the effect of a cyclic shift on the test vector of s. The test vector of the new sequence is

$$((1, 1, \ldots, 1) + (p - 1 + a_{p-1}, 0 + a_0, 1 + a_1, \ldots, p - 2 + a_{p-2})) \bmod p;$$

that is, the original test vector cyclically shifted plus a constant vector and all this modulo p. Clearly, this new test vector contains all elements of $[p]$ if and only if the original one does.

Applied to s, the flattening algorithm introduced in the last section leads, via site swaps and cyclic shifts, to a finite constant sequence or to a sequence of the form $m(m - 1) \cdots$. The test vector of the first kind of sequence corresponds to a permutation, and the test vector of the second one does not. Together with our considerations of the effects of site swaps and cyclic shifts on test vectors, this proves our result.   □

For a different proof of this result, see the proof of the respective result for multiplex juggling sequences on page 68 and the proof of [19], Theorem 2.

A straightforward consequence of the permutation test is the following result:

Vertical Shifts

Let $s = \{a_k\}_{k=0}^{p-1}$ be a sequence of nonnegative integers, let d be an integer that is greater than or equal to $-\min\{a_0, a_1, \ldots, a_{p-1}\}$, and let s' be the sequence $\{a_k + d\}_{k=0}^{p-1}$. Then, s is a juggling sequence if and only if s' is. We call the operation of transforming s into s' a *vertical shift of distance d*.

For example, we can add 5 to every element of the juggling sequence 441 to arrive at the juggling sequence 996, or we can subtract 1 from every element to arrive at the juggling sequence 330.

2.6.1 A Method to Construct All Juggling Sequences

The permutation test also provides an explicit way of constructing all b-ball juggling sequences of period p that can be easily implemented on a computer.

We start by listing all test vectors that correspond to permutations of $[p]$. Given one such test vector P, we reconstruct all b-ball juggling sequences that correspond to this test vector. For this, we calculate

$$P' = (P - (0, 1, 2, \ldots, p - 1)) \bmod p.$$

Both the sums of the entries of P and of the vector that gets subtracted from P to arrive at the new vector P' equal

$$\sum_{k=0}^{p-1} k.$$

This implies that the sum modulo p of the entries of P' is 0. Hence, the average of the new vector P' is a certain integer a. Furthermore, since all entries of P' are nonnegative integers less than or equal to $p - 1$, we also conclude that $0 \le a \le p - 1$. Let $b' = b - a$. Then, the different solutions to our problem are in one-to-one correspondence with the vectors of length p with nonnegative integer entries whose sum is b'. Explicitly, if Q is one such vector, then the entries of $P' + pQ$ form the associated b-ball juggling sequence of period p. In particular, if b' is negative, then the original test vector does not correspond to any of the juggling sequences we are after.

As an example, we calculate all 3-ball juggling sequences of period 3 in Table 2.1. There are 37 such sequences, but there is a lot of redundancy in this list. Up to cyclic shifts, there are exactly 13 sequences. These are highlighted. All of these 13 sequences are minimal with the exception of the sequence 333. If we actually wanted to turn our procedure into an efficient computer program, we would clearly want to include some subroutines that cut down on the number of calculations that need to be done. One such subroutine is suggested by the last three columns: If in the list of possible vectors for P' you find two that, up to cyclic shifts, are the same, then the juggling sequences that correspond to them also are identical up to cyclic shifts. We may therefore restrict ourselves to calculating the corresponding juggling sequences for only one of these vectors.

Deriving the 3-Ball Juggling Sequences of Period 3						
P(ermutation)	(0,1,2)	(2,0,1)	(1,2,0)	(1,0,2)	(2,1,0)	(0,2,1)
$P' =$ $(P - (0,1,2))$ mod 3	(0,0,0)	(1,1,1)	(2,2,2)	(1,2,0)	(2,0,1)	(0,1,2)
$b -$ average(P')	3	2	1	2	2	2
possibilities for the vector Q	(3,0,0) (0,3,0) (0,0,3) (2,1,0) (2,0,1) (0,2,1) (1,0,2) (0,1,2) (1,1,1)	(2,0,0) (0,2,0) (0,0,2) (1,1,0) (1,0,1) (0,1,1)	(1,0,0) (0,1,0) (0,0,1)	(2,0,0) (0,2,0) (0,0,2) (1,1,0) (1,0,1) (0,1,1)	(2,0,0) (0,2,0) (0,0,2) (1,1,0) (1,0,1) (0,1,1)	(2,0,0) (0,2,0) (0,0,2) (1,1,0) (1,0,1) (0,1,1)
juggling sequences $P' + 3Q$	**900** 090 009 **630** **603** 063 360 036 306 **333**	**711** 171 117 **441** 414 144	**522** 252 225	**720** **180** **126** **450** **420** **153**	027 018 612 045 042 351	702 801 261 504 204 135

TABLE 2.1. Deriving the 3-ball juggling sequences of period 3 from the six permutations of the numbers 0, 1, and 2. There are 37 such sequences. Up to cyclic shifts, there are 13 (typeset in bold letters).

2.6.2 Inverse of a Juggling Sequence

Starting with a juggling sequence, we draw its juggling diagram and observe that we can reflect the diagram through a vertical line in the plane to arrive at a juggling diagram of another juggling sequence. Of course, this new "inverse" juggling sequence is made up of exactly the same throws as the original one. However, it may not coincide with the original one, not even up to cyclic shifts. In the following, we will figure out an explicit formula for this "inverse."

Given a juggling sequence $s = \{a_k\}_{k=0}^{p-1}$, let us agree that we perform the a_k-throws exactly on the beats

$$\ldots, -2p + k, -p + k, 0 + k, p + k, 2p + k, \ldots.$$

We define a sequence $s' = \{b_k\}_{k=0}^{p-1}$, where b_k is the height of the throw in the original juggle ending on beat k. Then, the inverse of the original sequence s is the sequence

$$s' = \{c_k\}_{k=0}^{p-1} = \{b_{p-1-k}\}_{k=0}^{p-1}.$$

By definition, the inverse s' is a juggling sequence. Furthermore, it is clear that the inverse of s' is s.

As an example, let's consider the juggling sequence 56414. From its juggling diagram in Figure 2.14, we see that its inverse is 14645. Even using cyclic shifts, it is not possible to transform the first sequence into the second. Of course, up to cyclic shifts, the inverse of every juggling sequence of period 1 or 2 coincides with its inverse. The inverse of a juggling sequence $a_0 a_1 a_2$ of period 3 coincides up to cyclic shifts either with the sequence itself or with the sequence $a_2 a_1 a_0$. For example, among the 13 essentially different 3-ball juggling sequences of period 3 listed in Table 2.1, all except for two are their own inverses up to cyclic shifts. The two exceptions 630 and 603 are inverses of each other (up to cyclic shifts).

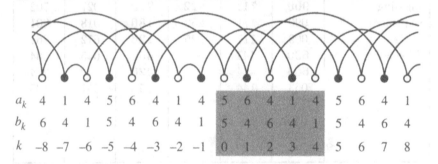

FIGURE 2.14. The inverse of the juggling sequence 56414 is the juggling sequence 14645.

To explicitly figure out in general what the elements $c_j = b_{p-1-j}$ of s' are, observe that there is exactly one $k \in [p]$ such that

$$j = (k + a_k) \bmod p.$$

This means that

$$j = \phi_s(k),$$

where ϕ_s is the permutation that corresponds to s. Hence,

$$b_{\phi_s(k)} = a_k,$$

which, of course, is equivalent to saying that

$$b_k = a_{\phi_s^{-1}(k)}.$$

Calculating the Inverse of 56414					
k	0	1	2	3	4
a_k	5	6	4	1	4
$\phi_s(k)$	0	2	1	4	3
$\phi_s^{-1}(k)$	0	2	1	4	3
$p-1-k$	4	3	2	1	0
$\phi_s^{-1}(p-1-k)$	3	4	1	2	9
$c_k = a_{\phi_s^{-1}(p-1-k)}$	1	4	6	4	5

TABLE 2.2. Starting with the juggling sequence 56414, the table lists the individual steps necessary to calculate its inverse, 14645.

We conclude that the entries of the inverse of s are of the form

$$c_k = a_{\phi_s^{-1}(p-1-k)}.$$

Table 2.2 summarizes the individual steps necessary to calculate the inverse of our sample sequence 56414.

Let us double-check that s' is in general really a juggling sequence. For this, we have to show that $\phi_{s'}$ is a permutation. We know that the entries of s can be written in the form

$$a_k = \phi_s(k) + pr_k - k,$$

where r_k is a certain integer. Hence,

$$\phi_{s'}(k)$$

$$= (c_k + k) \bmod p$$

$$= (a_{\phi_s^{-1}(p-1-k)} + k) \bmod p$$

$$= (\phi_s(\phi_s^{-1}(p-1-k)) + pr_{\phi_s^{-1}(p-1-k)} - \phi_s^{-1}(p-1-k) + k) \bmod p$$

$$= p - 1 - \phi_s^{-1}(p-1-k).$$

Clearly, since ϕ_s is a permutation, so is $\phi_{s'}$. We summarize the results of this subsection as follows:

The Inverse of a Juggling Sequence

Let $s = \{a_k\}_{k=0}^{p-1}$ be a juggling sequence, and let ϕ_s be its associated permutation. Then, its inverse is the juggling sequence

$$s' = \{a_{\phi_s^{-1}(p-1-k)}\}_{k=0}^{p-1}$$

and

$$\phi_{s'}(k) = p - 1 - \phi_s^{-1}(p - 1 - k)$$

for $k = 0, 1, \ldots, p - 1$.

2.6.3   Pick a Pattern Procedure

In [94], page 168, Martin Probert gives a recipe for constructing some of the b-ball juggling sequences of period p that boils down to the following. Start by making up a $p \times p$ matrix whose (i,j)th entry is $b - i + j$. This means that all entries on the main diagonal are bs and that every entry of the kth diagonal parallel to the main diagonal is $b + k$; see Figure 2.15.

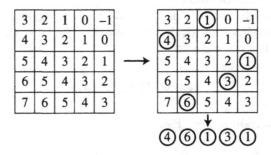

FIGURE 2.15. Probert's Pick a Pattern Procedure for constructing 3-ball juggling sequences of period 5.

Among the $p \times p$ entries of this matrix, highlight p nonnegative entries such that every row and column contains exactly one highlighted entry. Let a_k be the highlighted entry in column $k + 1$ for $k = 0, 1, \ldots, p - 1$. Then, $\{a_k\}_{k=0}^{p-1}$ is a b-ball juggling sequence of period p. Using our permutation test, it is very easy to verify that this procedure really works. Just add to every entry in the ith column of the matrix the number $i - 1$ and reduce modulo p. This gives a new matrix whose (i, j)th entry is

$$(b - i + j + i - 1) \bmod p = (b + j - 1) \bmod p.$$

This means that, given any row in the new matrix, all its entries are equal, and every integer from 0 to $p - 1$ is the common value of all entries in exactly one row. This implies that since the a_ks were chosen from different rows in the original matrix, the test vector

$$((a_0 + 0) \bmod p, (a_1 + 1) \bmod p, \dots, (a_{p-1} + p - 1) \bmod p)$$

will contain every integer from 0 to $p - 1$ exactly once. Hence, we are really dealing with a juggling sequence.

Concatenating two b-ball juggling sequences of periods p_1 and p_2 constructed like this yields a b-ball juggling sequence of period $p_1 + p_2$ that also arises in this manner. A quick glance at the example in Figure 2.16 should be enough to tell you why this is so.

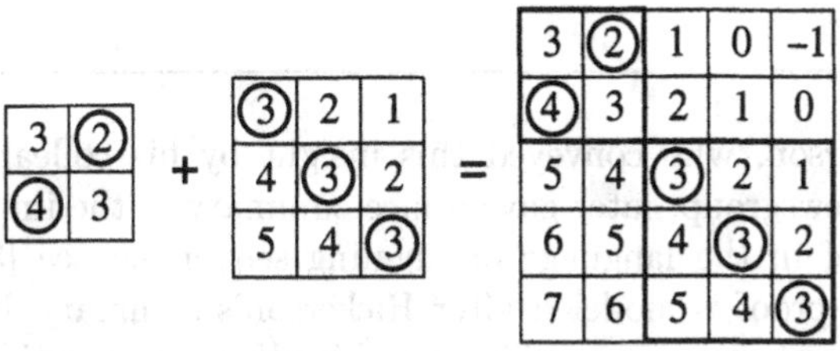

FIGURE 2.16. Joining two "picked" b-ball juggling sequences yields a new picked b-ball juggling sequence. In this example, the sequences 42 and 333 combine into the new sequence 42333.

How many b-ball juggling sequences of period p can be produced with this method? If $p \leq b + 1$, then there are no negative integers in the matrix that have to be avoided. In this case, we can choose any of the p entries in the first row. Following this, we can only choose among $p - 1$ entries in the second row, $p - 2$ in the third row, and so forth, respectively. Hence, the number of sequences is $p!$.

Now, let's assume that $p > b + 1$. If $j \leq p - (b + 1)$, then there are exactly $(b+1) + (j-1)$ nonnegative entries in row j, and if $j > p - (b+1)$, then there are exactly p nonnegative entries in row j. Now, we can argue as above, starting with the first row and working our way towards the last row, to conclude that there are a total of

$$(b+1)^{(p-b)} b(b-1)(b-2) \cdots 1 = b!(b+1)^{p-b}$$

juggling sequences that can be picked in this case.

In Subsection 2.8.2, we will see that the "picked" b-ball juggling sequences are exactly the so-called ground-state b-ball juggling sequences.

2.6.4 Converse of the Average Theorem

Suppose we are given a finite sequence of nonnegative integers whose average is an integer. We already noted that this does not imply that the sequence is a juggling sequence. Recently, the mathematical jugglers on rec.juggling got interested in figuring out whether there is always a re-arrangement of such a sequence that is a juggling sequence. After much discussion, it was reported (see [53]) that such a rearrangement always exists and that this result is a special case of a theorem about Abelian groups that was proved by Marshall Hall in 1952; see [50].

The "Converse" of the Average Theorem

Given a finite sequence of nonnegative integers whose average is an integer, there is a permutation of this sequence that is a juggling sequence.

Dean Hickerson, who conveyed this insight by his colleague Sherman Stein to the newsgroup, later gave a nice summary of the important steps in Hall's proof in the language of juggling sequences; see [54] and [55]. The following proof is modeled after Hickerson's summary. In particular, we use his sample sequences. The proof itself is constructive; that is, if you understand it, you will be able to rearrange any *qualifying* sequence into a juggling sequence. Here, a qualifying sequence is a finite sequence of nonnegative number whose average is an integer.

Proof. We prove the result for all qualifying sequences of a fixed period p. In the following, all additions and subtractions will be mod p. With this convention in place, it is clear that a finite sequence of nonnegative integers qualifies if and only if the sum of its elements is zero.

The main step of the argument consists in proving the following lemma:

Given a qualifying sequence that can be rearranged into a juggling sequence, replace any two of its elements by any two nonnegative integers such that the new sequence still qualifies. Then, the new sequence can also be rearranged into a juggling sequence.

Now, it is easy to see that the trivial juggling sequence $00\cdots 0$ of period p can be transformed into a given qualifying sequence of period p in no more than $p - 1$ steps in which we move from qualifying sequence to qualifying sequence using the 2-element substitutions specified in the lemma. Here is an example that can be generalized in a straightforward manner. Suppose that $p = 6$ and the qualifying sequence we are focusing on is 023355. Then, we can transform the trivial sequence 000000 into 023355 as follows:

$$000000 \rightarrow 024000 \rightarrow 023100 \rightarrow 023340 \rightarrow 023355.$$

Basically, we move from left to right such that on every step the first number from the left that does not coincide with the number in the respective position in our target sequence gets adjusted. Since we start with a juggling sequence, the lemma above now guarantees that all intermediate sequences, and therefore also our target sequence, can be rearranged into juggling sequences.

It remains to prove the lemma. In fact, up to some obvious reshuffling, it suffices to restrict ourselves to proving the lemma for the special case where the starting sequence is a juggling sequence. In doing this, it helps to keep a specific example in mind. We choose the juggling sequence 53052630 (period $p = 8$). We start with the following array:

$$
\begin{array}{cccccccc}
0 & 1 & 2 & 3 & 4 & 5 & 6 & 7 \\
5 & 3 & \mathbf{0} & 5 & \mathbf{2} & 6 & 3 & 0 \\
5 & 4 & 2 & 0 & 6 & 3 & 1 & 7
\end{array}
$$

Note that the third entry in each column is the sum (mod 8) of the first two entries. The first row consists of the integers from 0 to 7 in their natural order, and the second is the juggling sequence. Furthermore, since 53052630 is a juggling sequence, the last row is a permutation of the integers from 0 to 7.

Now, we decide to change the two numbers that are written in bold by two 1s. We may do this since the sum of the elements in the resulting sequence 53151630 is still zero.

We remove the two numbers in the second row that we want to change and write the two numbers that we want to replace them by at the right. We also move the corresponding two numbers in the third row to the right. We call the resulting array an *auxiliary array*.

$$
\begin{array}{ccccccccccc}
0 & 1 & 2 & 3 & 4 & 5 & 6 & 7 & & & \\
5 & 3 & - & 5 & - & 6 & 3 & 0 & & 1 & 1 \\
5 & 4 & - & 0 & - & 3 & 1 & 7 & & 2 & 6
\end{array}
$$

In this array, we are interested in six entries denoted $a_0, a_1, b_0, b_1, c_0,$ and c_1 as follows:

$$
\begin{array}{ccccccccccc}
0 & 1 & a_0 & 3 & a_1 & 5 & 6 & 7 & & & \\
5 & 3 & - & 5 & - & 6 & 3 & 0 & & b_0 & b_1 \\
5 & 4 & - & 0 & - & 3 & 1 & 7 & & c_0 & c_1
\end{array}
$$

In our example, $a_0 = 2$, $a_1 = 4$, and so forth. It is clear that

$$
a_0 + a_1 + b_0 + b_1 = c_0 + c_1 \tag{$*$}
$$

because the two numbers b_0 and b_1 have the same sum as the numbers they replace.

Now, we check if either of the sums $a_0 + b_0$ or $a_0 + b_1$ is equal to c_0 or c_1. Suppose, for example, that $a_0 + b_1 = c_0$. Then, it follows from equation $(*)$

that $a_1 + b_0 = c_1$. Consequently, we get a juggling sequence in row two by moving b_1 to column a_0 and b_0 to column a_1. We proceed in a similar way in the remaining three cases under consideration.

Assume that this first attempt at rearranging the sequence did not work. Let $x = c_0 - b_0$. Then, x is neither a_0 nor a_1 and, consequently, the second and third entries of column x are not empty. Let's call these entries b_2 and c_2, respectively. Now, we change our auxiliary array from

$$\begin{array}{ccccc} \dots & x & \dots & & \\ \dots & b_2 & \dots & b_0 & b_1 \\ \dots & c_2 & \dots & c_0 & c_1 \end{array} \quad \text{to} \quad \begin{array}{ccccc} \dots & x & \dots & & \\ \dots & b_0 & \dots & b_2 & b_1 \\ \dots & c_0 & \dots & c_1 & c_2 \end{array}$$

Note that $x + b_0 = c_0$. Also,

$$\begin{aligned} a_0 + a_1 + b_2 + b_1 &= a_0 + a_1 + b_0 + b_1 + b_2 - b_0 \\ &= c_0 + c_1 + b_2 - b_0 \\ &= c_1 + b_2 + c_0 - b_0 \\ &= c_1 + b_2 + x \\ &= c_1 + b_2 + c_2 - b_2 \\ &= c_1 + c_2; \end{aligned}$$

This means that equation (*) still holds for the new auxiliary array. In our example, we have $x = 1$, and the new auxiliary array is

0	1	2	3	4	5	6	7		
5	**1**	-	5	-	6	3	0	**3**	**1**
5	**2**	-	0	-	3	1	7	**6**	**4**

We relabel the four numbers that are sticking out of the array on the right b_0, b_1, c_0, and c_1, as in our first auxiliary array. Note that a_0, a_1, and b_1 have not changed. Now, things repeat; that is, we check if either of the sums of $a_0 + b_0$ or $a_0 + b_1$ is equal to c_0 or c_1. If this is the case, then we proceed as above to arrive at the rearrangement we are after. Otherwise, we calculate a new x and so forth. It remains to show that this process terminates in a finite number of steps. Before we do this, let's see what happens to our example if we proceed in this manner.

Check that we have to calculate a new x and that $x = 6 - 3 = 3$. We reshuffle to get the new auxiliary array

0	1	2	3	4	5	6	7		
5	1	-	**3**	-	6	3	0	**5**	**1**
5	2	-	**6**	-	3	1	7	**4**	**0**

We relabel again but find that we have to rearrange again. This time, $x = 4 - 5 = 7 \pmod 8$. The new auxiliary array is

0	1	2	3	4	5	6	7		
5	1	-	3	-	6	3	5	0	1
5	2	-	6	-	3	1	4	0	7

We have to rearrange again with $x = 0 - 0 = 0$.

0	1	2	3	4	5	6	7		
0	**1**	-	**3**	-	**6**	**3**	**5**	**5**	**1**
0	**2**	-	**6**	-	**3**	**1**	**4**	**7**	**5**

Finally, we find what we are looking for:

$$a_0 + b_0 = 2 + 5 = 7 = c_0$$

and consequently also

$$a_1 + b_1 = 4 + 1 = 5 = c_1.$$

We move $b_0 = 5$ to column two and $b_1 = 1$ to column four to arrive at the juggling sequence 01531635.

We still need to show that the algorithm described above always terminates after a finite number of steps. For this, it suffices to prove that the various xs that we come across in the course of our manipulations are all different. Then, since x can take on at most $p - 2$ different values, we are done.

The proof will be by contradiction. Let us assume that we come across the same x at least twice. The first time this happens, we change

...	x	...					...	x	...		
...	b_2	...	b_0	b_1	to	...	b_0	...	b_2	b_1
...	c_2	...	c_0	c_1		...	c_0	...	c_1	c_2

Let us call these two arrays A and B. The second time we come across the same x, we change

...	x	...					...	x	...		
...	b_2'	...	b_0'	b_1	to	...	b_0'	...	b_2'	b_1
...	c_2'	...	c_0'	c_1'		...	c_0'	...	c_1'	c_2'

We call these two arrays C and D. Note that the entry b_1 never changes. Let us focus on what happens after we have written down the auxiliary array B. We compute $x' = c_1 - b_2$ and move c_1 to the bottom of column x'. Since $x = c_2 - b_2$ and c_1 and c_2 are different, we conclude that x and x' are different. Furthermore, c_1 will stay at the bottom of column x' unless a subsequent value of x is also equal to x'. Since we are assuming that the value of x in the auxiliary arrays A and C is the first one to be repeated, x' cannot reoccur as a value of x until after the auxiliary array C. So, in auxiliary array C, the entry c_1 is still at the bottom of column x'. Because this is the first time that x shows up again, we can be sure that c_1 is not equal to c_1'. However, in the following, we will show that $c_1 = c_1'$, which is the contradiction we are after: Equation $(*)$ for the two auxiliary arrays on the left is

$$a_0 + a_1 + b_0 + b_1 = c_0 + c_0 \text{ and } a_0 + a_1 + b_0' + b_1 = c_0' + c_0'.$$

We subtract the second from the first to arrive at the equation

$$b_0 - b_0' = c_0 + c_1 - c_0' - c_1'.$$

Also, by definition,

$$x + b_0 = c_0 \text{ and } x + b_0' = c_0'.$$

Hence,

$$\begin{aligned} b_0 - b_0' &= (x + b_0) + c_1 - (x + b_0') - c_1' \\ &= b_0 - b_0' + c_1 - c_1', \end{aligned}$$

and, consequently, $c_1 = c_1'$. With this contradiction, we are done.   □

It seems like a straightforward exercise to turn this proof into a computer program that rearranges qualifying sequences into juggling sequences. However, nobody seems to have done this yet. It would be interesting to devise some algorithm that, upon input of a qualifying sequence, finds all rearrangements of this sequence into juggling sequences. Of course, a brute force approach would be to check all permutations of the sequence for jugglability. There should be a smarter way. In Subsection 2.6.6, we deal with a special case of this problem.

Another interesting problem would be to find some good estimates for the average number of rearrangements of juggling sequences of fixed period p into juggling sequences.

2.6.5 Scramblable Juggling Sequences

In the previous subsection, we convinced ourselves that every finite sequence of nonnegative integers whose average is an integer can be rearranged into a juggling sequence. In a number of articles on rec.juggling, it was pointed out that there are *scramblable juggling sequences*. These are juggling sequences that stay juggling sequences no matter how you rearrange their elements. Simple examples of such juggling sequences are 147 and 1999.

Using the permutation test, it is easy to characterize the scramblable juggling sequences. First, observe that the constant b-sequence of period p is scramblable. Now, in view of the permutation test, it is clear that if we add multiples of p to the individual elements of this trivial sequence, we also get a scramblable juggling sequence. In fact, we will see in a moment that every scramblable juggling sequence arises in this manner. For example, to construct 1999, start with the sequence 1111 (period 4) and add 8 (a multiple of the period) to the last three elements of this sequence.

Scramblable Juggling Sequences

A finite sequence of nonnegative integers is a scramblable juggling sequence of period p if and only if it is of the form $\{a_k p + c\}_{k=0}^{p-1}$, where c and the a_ks are nonnegative integers.

From what we said before, it is clear that every sequence of this special form is a scramblable juggling sequence.

Conversely, given a juggling sequence of period p that is scramblable, reduce all its elements mod p. Then, this reduced sequence is also a scramblable juggling sequence. If the reduced sequence is not constant, then it contains two elements a and b such that $a > b$. Scramble the reduced sequence such that a is the 0th and b is the $(a - b)$th element. Using the permutation test, it is easy to see that this new sequence is not jugglable because $a + 0 = b + (a - b)$, which is a contradiction to our assumption. Hence, the reduced sequence is constant, and the sequence we started with is of the special form under discussion.

Of course, every juggling sequence of period 1 or 2 is scramblable, and there are nonscramblable juggling sequences of period p for any $p > 2$.

2.6.6 Magic Juggling Sequences

A special case of one of the problems mentioned at the end of Subsection 2.6.4 is to enumerate the *magic juggling sequences*. A magic juggling sequence is a juggling sequence of some period p that contains every integer from 0 to $p-1$ exactly once. Here, the "magic" is a similar kind of magic as the magic in "magic squares." It is clear that the inverse and every cyclic shift of a magic juggling sequence is also a magic juggling sequence.

The average of a magic juggling sequence of period p is

$$\frac{\sum_{k=0}^{p-1} k}{p} = \frac{p-1}{2}.$$

This means that, by the Converse of the Average Theorem (see Subsection 2.6.4), magic juggling sequences exist if and only if the period p is odd. Table 2.3 lists the simplest magic juggling sequences.

Here is an easy way to generate magic juggling sequences.

Some Magic Juggling Sequences

Let p and q be two positive integers such that p is odd, $q > 1$, and p is relatively prime to both q and $q - 1$. Then,

$$\{(q-1)k \bmod p\}_{k=0}^{p-1}$$

is a magic juggling sequence.

Jon Stadler pointed this out in a post to `rec.juggling`; see [126]. This result is an easy consequence of the permutation test and the following basic number-theoretic result:

Given two positive integers p and q that are relatively prime, the function

$$[p] \to [p] : k \mapsto kq \bmod p$$

is a permutation of $[p]$.

Since p is odd, we can always choose $q = 2$. In doing so, we find that

$$012 \cdots (p-1)$$

is a magic juggling sequence (this is real magic!). Since the construction of these sequences works mod p, choosing q among the integers in the range from 2 to $p - 1$ will yield all the different magic juggling sequences that can be generated in this way for the given choice of p. In fact, we will get different juggling sequences for different choices of q within this range. For example, let $p = 7$. Then, choosing q to be 2, 3, 4, 5, 6 yields the sequences

$$0123456, 0246135, 0362514, 0415263, 0531642,$$

respectively.

Magic Juggling Sequences up to Period 7						
balls	0	1	2	3		
sequences	0	012	01234	0123456	0246135	0362514	0461253
			02413	0135264	0245163	0413562	0512463
			03142	0142635	0263145	0415263	0526134
				0236415	0315246	0416235	0531642
				0241536	0346152	0425613	

TABLE 2.3. The different (up to cyclic shifts) magic juggling sequences that can be juggled with 0, 1, 2, and 3 balls.

To choose $q = 2$ was one obvious choice to get a general result for all possible choices of p. Of course, a second obvious choice is $q = p - 1$. The corresponding magic juggling sequence is

$$0(p - 2)(p - 4) \cdots 1(p - 1)(p - 3) \cdots 2,$$

which can be cyclically shifted into the interesting magic juggling sequence

$$(p - 2)(p - 4) \cdots 1(p - 1)(p - 3) \cdots 0$$

in which we first juggle all odd numbers among the numbers from 0 to $p-1$ and then the even numbers (in their reverse natural orders).

2.7   How Many Ways to Juggle?

"How many patterns are there?" is a question that jugglers get asked very often. Of course, since b is a juggling sequence for every nonnegative integer b, it is clear that the answer is "infinitely many."

However, it still makes sense to ask for the number of juggling sequences that are distinguished in some way. The three most natural parameters used to define distinguished classes of juggling sequences are:

- the number of balls used to juggle a juggling sequence;

- the period of a juggling sequence;

- the maximum height of a throw in a juggling sequence.

So, how many juggling sequences do we get if we fix or limit one, two, or all three parameters?

If we only fix the number of balls, or the period, or a maximum throw height, the resulting class of juggling sequences will still be infinite, except for some trivial exceptions.

Fixing the period p and a maximum throw height h yields a finite class of juggling sequences. Clearly, there are no more than $(h+1)^p$ such sequences. As a consequence of the Average Theorem B1 (see page 15), all sequences in this class are juggled with at most h balls, and the only h-ball juggling sequence in this class consists of hs only. If b is a nonnegative integer such that $pb \le h$, then again as a consequence of the Average Theorem B1, all b-ball juggling sequences of period p are contained in this class.

Apart from some trivial exceptions, fixing the number of balls and maximum throw height will give an infinite class of juggling sequences. However, we will see in Section 2.8 that a lot of interesting results can be proved about the b-ball juggling sequences that consist of throws up to a certain fixed maximum height. This includes, among other things, very effective ways of listing all such sequences of a given period p by finding the number of loops in a certain graph. Of course, this is exactly what we are interested in when limiting all three parameters.

2.7.1   Juggling Cards

In Subsection 2.6.1, we described a way to construct all b-ball juggling sequences of period p starting with the permutations of the integers from 0 to $p-1$. It is clear that every permutation only yields a finite number of such sequences and that therefore the number of all b-ball juggling sequences of period p is finite. In [19], Joe Buhler, Ronald Graham, Colin Wright, and David Eisenbud determined the number of all such sequences by counting the number of juggling sequences that correspond to the different permutations and then summing over all numbers obtained in this way; see also [17] and [18]. They also reported that Adam Chalcraft, Jeremy Kahn, Richard Stanley, and Walter Stromquist each independently from one another came up with the same result using a different approach. In [17], Joe Buhler and Ronald Graham generalized the results in [19] to a result about posets. This result was reproved by Einar Steingrimsson in [130] and generalized even further. Finally, in [36], Richard Ehrenborg and Margaret Readdy generalized the results in [19] in a number of different ways that we will report on in this section and in Sections 3.2 and 3.3. In particular, we will follow the "juggling cards" idea in this paper to calculate the number of juggling sequences we are interested in. Note that in [36] juggling sequences are defined to be "juggling sequences without 0-throws." However, the arguments in this paper can be adapted in a straightforward manner to apply to the setting we are interested in, and we will state the results in [36] in terms of juggling sequences as we defined them above—that is, with 0-throws.

Have a look at the four different cards in Figure 2.17.

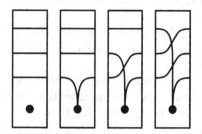

FIGURE 2.17. The four juggling cards C_0, C_1, C_2, C_3 that can be used to generate all juggling sequences that are juggled with at most three balls.

Imagine a deck of cards made up of infinitely many copies of each of these four cards. Shuffle the deck, randomly pick p cards from the deck, and place them in front of you as the gray cards in Figure 2.18. Repeat this arrangement of cards infinitely often to the left and right as indicated and you arrive at a juggling diagram of a juggling sequence of period p and number of balls at most 3. In fact, it turns out that every such juggling sequence corresponds to exactly one ordered set of p cards from the deck.

Of course, once you have constructed a juggling diagram using the cards, it is clear which sequence you are dealing with.

Note that if you do not use a copy of the last card in Figure 2.17, you will end up with (at least) one continuous horizontal line on top. This just means that the juggling sequence uses no more than two balls. In fact, the highest index of the cards used equals the number of balls used.

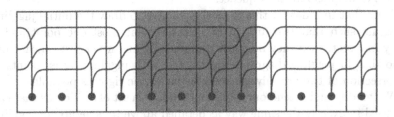

FIGURE 2.18. Any p juggling cards in random order will generate a juggling sequence of period p. In this example, $p = 4$ and the sequence is 4053.

Given a juggling sequence, how do you figure out how to construct its juggling diagram using the cards? Here is what you have to do. Suppose we are dealing with our favorite sequence 441. Draw the juggling diagram in the usual manner; see Figure 2.19.

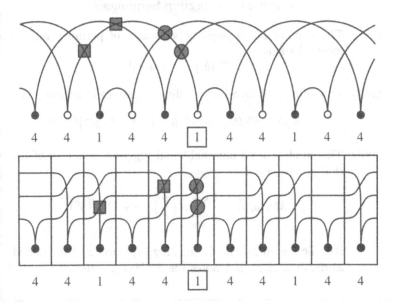

FIGURE 2.19. Representing the juggling sequence 441 in terms of juggling cards.

To every number in the sequence there corresponds exactly one card. If the number is a 0, choose a copy of card C_0 and place it above the 0. Choose any of the numbers that are not 0, say the 1 (boxed in the diagram).

From the circle that indicates the corresponding beat, trace backwards on the incoming arc and inspect the crossings with other arcs. When we cross another arc, we either cross from the inside to the outside or in the other direction. In the first case, we mark the crossing by a gray circle, otherwise by a gray square. Count the number c of gray circles. In our example, there are $c = 2$ such circles. Choose card C_c and place it above the 1. Repeat for all other numbers in the sequence.

Note that in order for this to work you have to draw the initial juggling diagram such that two arcs in it intersect in at most one point and if they intersect, then they intersect transversally and never just touch. Note also that in juggling diagrams constructed with juggling cards, points of intersection of two arcs always occur as far to the right as possible.

We modify and extend our set of four cards to a set of $b+1$ cards that can be used in exactly the same way as detailed above to generate all juggling sequences of period p and at most b balls. Card C_0 consists of b horizontal paths. On card C_i, $i \geq 1$, starting from the left, the ith path counted from below descends to the circle marking the beat and then continues as the first path on the right. Again starting from the right, the jth path is horizontal if $j > i$ and rises and continues as the $(j + 1)$th path if $j < i$.

Numbers of Juggling Sequences

(N1) The number of all juggling sequences of period p and at most b balls is
$$S^{\leq}(b, p) = (b + 1)^p.$$

(N2) The number of all b-ball juggling sequences of period p is
$$S(b, p) = S^{\leq}(b, p) - S^{\leq}(b - 1, p) = (b + 1)^p - b^p.$$

(N3) The number of all minimal b-ball juggling sequences of period p, with $b \geq 1$, is
$$MS(b, p) = \frac{1}{p} \sum_{d \mid p} \mu\left(\frac{p}{d}\right)\left((b + 1)^d - b^d\right)$$

if cyclic permutations of a juggling sequence are not counted as distinct. Here, μ denotes the *Möbius function*.[2]

---

[2] *Möbius function* μ. If n is a positive integer, then

$$\mu(n) = \begin{cases} 1 & \text{if } n = 1 \text{ or if } n \text{ has an even number of distinct prime factors,} \\ -1 & \text{if } n \text{ has an odd number of distinct prime factors,} \\ 0 & \text{if } n \text{ has repeated prime factors.} \end{cases}$$

The first two results follow immediately from our discussion above. Note that in $S(b, p)$ the cyclic permutations of any juggling sequence are counted as distinct, which gives a lot of redundancy. Also, juggling sequences that are not minimal are counted. For example, 333 is one of the sequences that is counted by $S(3,3)$. For a complete list of all 3-ball juggling sequences of period 3 counted by $S(3,3)$, see Table 2.1.

The number $MS(b, p)$ in N3, which avoids these redundancies, is probably of greater interest to jugglers. Here is how it is calculated. If d is a divisor of the period p, then each minimal juggling sequence of period d gives rise to exactly d sequences of period p. Therefore,

$$S(b, p) = (b + 1)^p - b^p = \sum_{d|p} d MS(b, d).$$

The expression for $MS(b, p)$ now follows by *Möbius inversion*.[3]

For small values of p, we calculate

$$MS(b, 1) = 1,$$
$$MS(b, 2) = b,$$
$$MS(b, 3) = b(b + 1),$$
$$MS(b, 4) = b(b^2 + b + 1),$$
$$MS(b, 5) = b(b^3 + 2b^2 + 2b + 1),$$

and, if the period p is a prime number, then the formula for $MS(b, p)$ reduces to

$$MS(b, p) = \frac{1}{p}((b + 1)^p - b^p - 1).$$

It is also easy to count those among the sequences considered in Results N1, N2, and N3 that do not feature 0-throws. Just note that if we remove card C_0 from our decks of cards, we will never get sequences with 0-throws. This implies, for example, that the number of b-ball juggling sequences of period p that do not feature 0-throws is $b^p - (b-1)^p$ for $b \geq 1$.

For example, $\mu(20) = \mu(5 \cdot 2 \cdot 2) = 0$ because 2 occurs repeatedly as a prime factor of 20.

[3] *Möbius inversion.* If f and g are functions defined for every positive integer p such that

$$f(p) = \sum_{d|p} g(d),$$

then

$$g(p) = \sum_{d|p} \mu\left(\frac{p}{d}\right) f(d).$$

See [51] for a proof of this result.

2.7.2 Weights of Juggling Sequences

In [36], the considerations above are also used to prove a number of interesting results about the affine Weyl group $\tilde{A}_{p-1}$. In the following, we give a summary of these results.

Given a juggling sequence s of period p and its corresponding juggling diagram made up from juggling cards, let $cross(s)$ be the number of points of intersection of arcs in the juggling diagram, different from the beat points, on p juggling cards making up a period. For example, $cross(4053) = 5$ and $cross(441) = 4$; see Figures 2.18 and 2.19. The *weight of* s is the formal expression $q^{cross(s)}$.

The Sum of the Weights

The sum of the weights of all juggling sequences of period p, at most $b \geq 1$ balls, and no 0-throws is equal to

$$\left(\sum_{i=0}^{b-1} q^i \right)^p .$$

The sum of the weights of all juggling sequences of period p and at most $b \geq 1$ balls is equal to

$$\left(1 + \sum_{i=0}^{b-1} q^i \right)^p .$$

This result, the first part of which is Theorem 1.1 in [36], follows immediately from our disscussion above and the fact that the sum of the weights of the respective b or $b+1$ cards used for constructing all juggling sequences under consideration is the sum in brackets.

Starting with the juggling sequence $s = \{a_i\}_{i=0}^{p-1}$, we now define a permutation of the integers by setting

$$\mathbf{Z} \to \mathbf{Z} : i \mapsto a_{i \bmod p} + i - b,$$

where b is the number of balls juggled. Then, it is easily checked that the set of all permutations that can be constructed in this manner is a group under composition. In fact, it turns out that this group is the affine Weyl group $\tilde{A}_{p-1}$. Two juggling sequences of period p correspond to the same element of $\tilde{A}_{p-1}$ if and only if one is the vertical shift of the other; see page 23 for a definition of the term vertical shift.

Clearly, $\tilde{A}_0$ is the one-element group. For $p \geq 2$, it can be shown that $\tilde{A}_{p-1}$ is generated by the *simple reflections*

$$t_k : \mathbf{Z} \to \mathbf{Z} : i \mapsto \begin{cases} i+1 & \text{for} \quad i \bmod p = k \bmod p, \\ i-1 & \text{for} \quad i \bmod p = (k+1) \bmod p, \\ i & \text{for} \quad i \bmod p \neq k, (k+1) \bmod p, \end{cases}$$

for $k = 0, 1, \ldots, p-1$.

Given an element σ of $\tilde{A}_{p-1}$, let $length(\sigma)$ be the smallest integer such that σ can be written as the product of this number of simple reflections. At the level of juggling sequences, the simple reflections translate into site swaps as follows. For sufficiently large b, consider the two b-ball juggling sequences that correspond to the two elements σ and σt_i, $0 \leq i \leq p-2$. Then, the juggling sequence corresponding to σt_i can be constructed from the juggling sequence corresponding to σ by a site swap of beats i and $i+1$. Also, the simple reflection t_{p-1} corresponds to a cyclic shift, followed by a site swap of beats 0 and 1, followed by a reverse cyclic shift.

The Affine Weyl Group $\tilde{A}_{p-1}$

If σ is the element of $\tilde{A}_{p-1}$, $p \geq 2$, corresponding to a b-ball juggling sequence s of period p without 0-throws, then

$$(b-1)p - length(\sigma) = cross(s).$$

The Poincaré series of $\tilde{A}_{p-1}$ is

$$\sum_{\sigma \in \tilde{A}_{p-1}} q^{length(\sigma)} = \frac{1 - q^p}{(1-q)^p}.$$

For a simple proof of this result that is based on the sum-of-weights result above, see [36] Theorem 4.2 and Corollary 4.3.

In [36], the juggling card idea is also adapted to enumerate the b-ball multiplex juggling sequences of height h and to calculate the sum of weights of such juggling sequences in terms of Gaussian coefficients. Furthermore, a number of nice interpretations of well-known combinatorial objects in terms of juggling sequences are described, and some slick proofs for important results involving these objects are given within the framework of juggling sequences. We summarize these results in Section 3.3.

2.8   Juggling States and State Graphs

Let's assume you are juggling some 3-ball juggling sequence. After a couple of minutes, you get bored and want to change your pattern without stopping the juggle. On the next beat, you want to do something different. What are your options? The answer to this question comes in the form of a juggling state. Remember that you are in between beats. Let's say balls 1, 2, and 3 are scheduled to land in 2, 3, and 5 beats from now. Then, the current *juggling state* or *landing schedule* is the following string of 0s and 1s:

01101

This juggling state says that there is no, 1, 1, no, 1, no, no, no,..., ball expected to land on beats 1, 2, 3, 4, 5, 6, ..., respectively, from now on. This means that we cannot do anything on the next beat because there won't be any ball to do anything with. In other words, you will perform a 0-throw on the next beat. Having done this, you are now faced with the juggling state

1101

which arises from the one above by canceling the first 0. On the next beat a ball will land and you have to do something with it. A 0-throw is not an option, a 1-throw would lead to a collision, a 2-throw is a possibility (too boring though), a 3-throw would result in a collision, any higher throw is possible, and you decide on a 6-throw. So, to construct the next juggling state, cancel the leading 1 from the last one, append two zeros, and, finally, append a 1 in position 6, which accounts for the fact that the ball you just tossed will land in 6 beats from now:

101001

Your options on the next toss are a 1-throw, a 3-throw, a 4-throw, and anything higher than 5. You can continue like this and beat by beat make up a juggling sequence as you go along. Of course, you can do the same with any number of balls. Admittedly, you would have to be a fairly quick thinker/juggler if you wanted to do all this in real time.

2.8.1   State Graphs

In general, a *b-ball juggling state of height h*, with $0 \le b \le h$ and $h \ge 1$, is a string consisting of b 1s and $h - b$ 0s. Clearly, there are

$$Vert(b,h) = \binom{h}{b} = \frac{h!}{b!(h-b)!}$$

such juggling states. These juggling states form the vertices of the *b-ball state graph of height h*. The edges of this graph are directed and are labeled

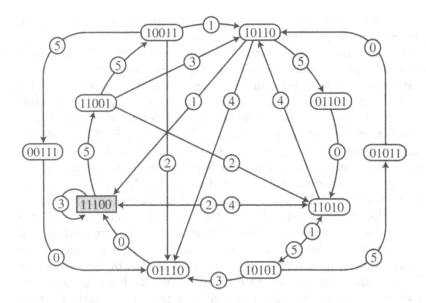

FIGURE 2.20. The 3-ball state graph of height 5.

with one of the integers from 0 to h. These edges are constructed as follows. Given a juggling state s whose first digit is a 0, move this 0 to the end of s to arrive at a second juggling state s'. Connect s and s' by an edge labeled with a 0 that originates at s and points at s'. Given a juggling state s whose first digit is a 1, cancel this leading 1 from s and append a 0 to arrive at an intermediate string of numbers. Replace any one of the 0s in this intermediate string, say in position i, by a 1 to arrive at a new juggling state s'. Connect s and s' by an edge labeled with the number i that originates at s and points at s'. This means that there are $h - b + 1$ edges originating at a state with a leading 1 and just one edge with a leading 0. Similarly, there are $b + 1$ edges pointing at a state that ends with a 0 and only one edge pointing at a state that ends with a 1. Figure 2.20 shows the 3-ball state graph of height 5.

We are now ready to state the relationship between juggling sequences and state graphs.

Juggling Sequences and Loops in State Graphs

The b-ball juggling sequences of height at most h, with $0 \le b \le h$ and $h \ge 1$, correspond exactly to the directed closed paths in the b-ball state graph of height h that start and end at the same state and contain at least one vertex and one edge.

Let's convince ourselves that this is really the case. From the considerations at the beginning of this section, it is clear that one of the directed paths under consideration translates into exactly one of the juggling sequences under consideration and how this translation works.

It remains to show that—and how—one of the juggling sequences under consideration translates into exactly one of the paths under consideration. Let's consider our favorite example, 441, and deduce what path it corresponds to in the state graph featured in Figure 2.20. Start by drawing its juggling diagram; see Figure 2.21. A vertical line through a point in between two beats intersects every one of the orbits in the juggling diagram exactly once. In our example, we are dealing with three orbits corresponding to the three balls. The three points of intersection of a vertical line before the two beats indexed 4 and 4 are marked by three triangles. Now, we just have to follow the orbits starting from these distinguished points to figure out how many more beats it will take the different balls to land (starting the count from the moment in time marked by the vertical line). The corresponding numbers then translate into the juggling state at this particular point in time. The triangles correspond to the state 11100, the squares to the state 11010, and the gray circles to the state 10110. This means that the path in the state graph above that corresponds to 441 is

11100(triangles) – 4 – 11010(squares) – 4 – 10110(circles) – 1 – 11100.

Of course, the reconstruction of the directed path in the state graph in this example generalizes in a straightforward manner to the general case. Furthermore, the periodic nature of juggling sequences and diagrams guarantees that the directed paths that can be reconstructed in this manner start and end at the same state and are uniquely determined.

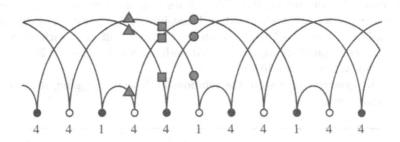

FIGURE 2.21. The balls corresponding to the three triangles will land in 1, 2, and 3 beats time. Hence, the corresponding juggling state is 11100.

In the following, we will often say that a juggling sequence originates at a state, visits a state, and so forth if the path that corresponds to it has this property.

2.8.2   Ground-State and Excited-State Sequences

Note that there is only one loop-edge in the b-ball state graph of height h; that is, an edge that originates and points at the same state. This distinguished edge is labeled with the number b, and the state with which it is associated is called the *ground state*. The ground state starts with b 1s and ends with $h - b$ 0s. In Figure 2.20, this state is marked by a gray rectangular box. The shortest closed path in any state graph starts and ends at the ground state and consists of the only loop-edge in the graph. The corresponding juggling sequence is the basic one-element juggling sequence b. All states different from the ground state are called *excited states*.

Here are a number of sample juggling sequences that originate and end at the ground state in Figure 2.20:

$$3 \quad 42 \quad 342 \quad 531 \quad 5340 \quad 55500$$

Note that any two juggling sequences that originate at the same state can be concatenated to form another juggling sequence that originates at the same state. For example, the juggling sequences 3 and 42 in the list above can be combined into the sequence 342. In fact, from what we said before, it is clear that two juggling sequences can be concatenated into a new juggling sequence if and only if both originate at the same state of some state graph. Similarly, any juggling sequence that visits a certain state can be extended to a longer juggling sequence by "splicing in" a juggling sequence that starts at this state. For example, the juggling sequence 55140, which orginates at the ground state, visits the state 10110 after the 1, and the juggling sequence 50253 starts at this state. After splicing the second sequence into the first sequence after the 1, we are left with the juggling sequence

$$5515025340.$$

Depending on whether or not a juggling sequence originates at the ground state, it is called a *ground-state* or *excited-state (juggling) sequence*, respectively. Note that 441 is a ground-state sequence, whereas 414 and 144 are not. In particular, a juggling sequence s is ground-state if and only if it is possible to go straight into it from juggling the basic one-element b-ball juggling sequence b and also if and only if you can go straight into juggling this basic sequence after juggling s. For example, we can juggle the 3-ball cascade and then go straight into 441, repeat this sequence three times, and then continue juggling the cascade; that is, juggle

$$\cdots 33334414414413333 \cdots .$$

Numbers of Ground- and Excited-State Sequences

(E1) The ground-state b-ball juggling sequences of period p are exactly those that can be constructed with Probert's Pick a Pattern Procedure described in Subsection 2.6.3. The number of such sequences is

$$E(b,p) = \begin{cases} p! & \text{if } p \leq b+1, \\ b!(b+1)^{p-b} & \text{otherwise.} \end{cases}$$

(E2) The number of excited-state b-ball juggling sequences of period p is

$$S(b,p) - E(b,p) = (b+1)^p - b^p - E(b,p).$$

See Section 2.7 for more information about the number $S(b,p)$ of b-ball juggling sequences of period p.

Remember that all constant b-ball juggling sequences can be picked and that any two picked b-ball juggling sequences can be concatenated into a new picked b-ball juggling sequence. This implies that the picked juggling sequences are all ground-state. Furthermore, we already showed in Subsection 2.6.3 that $E(b,p)$ is the number of picked b-ball juggling sequences of period p. It remains to show that there are no more ground-state b-ball juggling sequences of period p. To figure out what is happening, let us consider the special case $b = 3$, $p = 5$ illustrated in Figure 2.22.

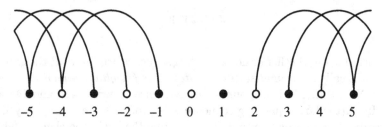

FIGURE 2.22. The setup for counting the number of 3-ball ground-state juggling sequences of period 5.

We start by juggling the basic 3-ball pattern. From beat -3 to beat 1, we juggle the five throws of one of the ground-state patterns we are after and then continue with the basic 3-ball pattern. In our example, $p > b+1$.

We have to account for the possible throws on the p beats

$$1, 0, -1, -2, -3.$$

On beat 1, we can choose among the $b + 1 = 4$ different throw heights 0, 1, 2, and 3. Following this, we have exactly $b + 1 = 4$ choices on beat 0. Continuing to argue like this, we see that there are $b = 3$, $b - 1 = 2$, and $b - 2 = 1$ choices on beats -1, -2, and -3, respectively. This gives a total of $b!(b + 1)^{p-b} = 3!4^2 = 96$ juggling sequences. This specific example generalizes in a straightforward manner to account for the general case where $p > b + 1$. The other case is dealt with in a similar fashion. The second part of this result is an immediate consequence of the first part and the explicit formula for the total number of b-ball juggling sequences of period p derived in Section 2.7.

2.8.3 Throws from States

Here is another neat result about juggling sequences that first popped up on the newsgroup rec.juggling.

States Determine Throws

If s and s' are two b-ball juggling sequences of height at most h that visit the same states in the b-ball state graph of height h the same number of times each, then s is a permutation of s'. More precisely, let d_0 be the total number of 0s in the first position of all states visited, and let d_i be the total number of 1s in the ith position of all states visited. Here, the 0s and 1s in a certain position of a state are counted as many times as the state is visited. Then, d_0 is the number of 0-throws, d_h is the number of h-throws, and $d_i - d_{i+1}$ is the number of i-throws for $1 \le i \le h-1$ in both s and s'.

Of course, this result is trivially true if s and s' correspond to the same loop in the state graph; that is, if they coincide up to cyclic shifts. As a nontrivial example, consider the two ground-state 3-ball juggling sequences of height 6 found by Willem Oudshoorn and optimized by Johannes Waldmann; see [89] and [142].

$$52660150530$$
$$53505166020$$

Figure 2.23 shows that one is a noncyclic permutation of the other and that the paths these two sequences correspond to in the 3-ball state graph of height 6 visit the same states exactly once.

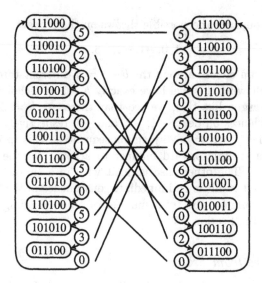

FIGURE 2.23. Two paths in the 3-ball state graph of height 6 that visit the same states the same number of times.

Of course, the result above, as we have stated it, contains its own proof in that we specify exactly how the elements of a juggling sequence can be reconstructed from the list of states it visits. Originally, the first part of the result was what people were trying to prove and the second part was the proof by Johannes Waldmann, whose proof, in turn, is a streamlined version of the original proof by Willem Oudshoorn; see [142] and [88], respectively. In the example depicted in Figure 2.23,

$$d_0 = 3, d_1 = 8, d_2 = 7, d_3 = 6, d_4 = 5, d_5 = 5, \text{ and } d_6 = 2.$$

Verify that our sample sequences indeed contain $3 = d_0$ 0-throws, $2 = d_6$ 6-throws, $1 = d_1 - d_2$ 1-throw, and so forth.

2.8.4   Prime Juggling Sequences and Loops

We call a loop in a state graph or one of its corresponding juggling sequences *prime* if it does not visit any state more than once.

Every juggling sequence can be decomposed into a number of prime juggling sequences. We illustrate what we mean by this by considering the ground-state 3-ball juggling sequence of height 5,

$$55150440.$$

You can check in the state graph in Figure 2.20 that this sequence visits the state 10110 twice. Starting from this state, we can juggle the juggling (sub-)sequence 504. This means that the sequence can be decomposed into

The Prime 3-Ball Juggling Sequences of Height at Most 5							
3	42	441	4440	45501	455040	5255040	55150530
	4 51 2	522	4530	52440	525501	5350530	
		531	5241	52530	551502		
		4 450 2	5340	53502			
			5511	55140			
			5520	55500			
				4 55050 2			

TABLE 2.4. The 26 prime 3-ball juggling sequences of maximum height 5 up to cyclic shifts. All except two of these sequences are ground-state sequences.

the prime juggling sequences

$$55140 \text{ and } 504.$$

   The period of a prime juggling sequence of height at most h is limited by the number of vertices in the b-ball state graph of height h and, consequently, there are only a finite number of such juggling sequences. Table 2.4 lists the 26 prime 3-ball juggling sequences of height at most 5 up to cyclic shifts. Except for the sequences 450 and 55050, all sequences in the table are ground-state sequences. The excited-state sequence 450 is preceded by the number 4. This means that the state 11010 at which it originates can be reached from the ground state via the edge labeled with a 4. Also, after you have juggled this sequence a number of times, you can return to the ground state by juggling a 2-throw. This is indicated by the 2 following the sequence in the table. The table was generated using the free *JuggleAnim Java Juggling Animator* by Jack Boyce; see Section 1.3 for more details about this program. The longest sequence corresponds to the circle-shaped path in the 3-ball state graph of height 5 depicted in Figure 2.20.
   A *subgraph* of a state graph consists of a number of its vertices and edges such that, for every one of the edges in the subgraph, the vertex at which it originates and the vertex it points at are contained in the subgraph. Consider the subgraph of the b-ball state graph of height h consisting of all its vertices and all edges labeled 0 or h. In this subgraph, any state is the origin of exactly one of its edges and the end point of exactly one of its edges. Given a state that starts with a 0 or 1, the edge that originates at it is labeled 0 or h, respectively. Similarly, given a state that ends with a 0 or 1, the edge that points at it is labeled 0 or h, respectively. This implies that the subgraph is the disjoint union of prime loops, which we will call the *necklaces* of both the subgraph and the state graph. The states contained in a necklace are precisely the cyclic shifts of any of the states contained in it. This means that a necklace has length at most h. For example, the necklace that contains the ground state has length h if $1 \le b \le h - 1$. In general, we conclude that the b-ball state graph of height h contains at

least

$$\frac{Vert(b,h)}{h} = \frac{\binom{h}{b}}{h}$$

necklaces. In fact, it is possible to say even more.

Number of Necklaces

The number of necklaces in the b-ball state graph of height h is

$$Neck(b,h) = \frac{1}{h} \sum_{d \mid gcd(b,h)} \phi(d) \binom{h/d}{b/d},$$

for $1 \le b \le h - 1$. Here, $gcd(b,h)$ is the greatest common divisor of b and h and ϕ is the *Euler function*.[4] Furthermore,

$$Neck(0,h) = Neck(h,h) = 1$$

for $h \ge 1$, and, in general,

$$Neck(b,h) = Neck(h-b,h).$$

The formula for $Neck(b,h)$ can be found in [124], Sequence A047996. For the following special values of b and h, with $b < h$, this formula simplifies considerably:

$$Neck(1,h) = 1;$$
$$Neck(2,h) = \begin{cases} \frac{h}{2} & \text{if } h \text{ is even,} \\ \frac{h-1}{2} & \text{if } h \text{ is odd;} \end{cases}$$
$$Neck(b,h) = \frac{Vert(b,h)}{h} = \frac{\binom{h}{b}}{h} \text{ if } gcd(b,h) = 1.$$

We call a prime loop in the state graph *trivial* if it is one of the necklaces or the one-edge loop. We call a prime loop *maximal* if its length is maximal among the lengths of prime loops in the state graph. Here are a couple of interesting questions that can be asked about prime loops in a given state graph:

---

[4] *Euler function ϕ.* If n is a positive integer greater than 1, then $\phi(n)$ is the number of elements in the set $\{1, 2, \ldots, n - 1\}$ that are relatively prime to n. Furthermore, $\phi(1) = 1$. If the decomposition of n into prime factors contains the distinct prime factors $p_1, p_2, \ldots, p_r$, then

$$\phi(n) = n \prod_{i=i}^{r} \left(1 - \frac{1}{p_i}\right).$$

- What is the length $MP(b, h)$ of a maximal prime loop in the b-ball state graph of height h?

- How many (maximal) prime loops are there?

We do not know the answers to these questions. However, we will derive an upper bound for $MP(b, h)$ that is very close to the actual value in the known cases.

The 0-ball state graph of any positive height contains only one vertex and one edge. So does the h-ball state graph of positive height h. Hence,

$$MP(0, h) = MP(h, h) = 1$$

for $h \geq 1$.

Let $1 \leq b \leq h - 1$. Then, there is at least one necklace of length h in the b-ball state graph of height h, and the number of vertices in the state graph is an absolute upper limit for $MP(b, h)$. Therefore,

$$h \leq MP(b, h) \leq \binom{h}{b}.$$

In particular, since the 1-ball state graph of height h and the $(h - 1)$-ball state graph of height h contain only one necklace each, we find that

$$MP(1, h) = MP(h - 1, h) = h.$$

It is easy to check that

$$MP(2, 4) = 4.$$

Let $2 \leq b \leq h - 2$ and $(b, h) \neq (2, 4)$. Then, a maximal prime loop is nontrivial. To see this, we follow the necklace of length h that contains the ground state

$$\underbrace{11 \cdots 1}_{b} \underbrace{00 \cdots 0}_{h-b}$$

to the state

$$1 \underbrace{00 \cdots 0}_{h-b} \underbrace{11 \cdots 1}_{b-1},$$

and then juggle an $(h - b - 1)$-throw to arrive at state

$$\underbrace{00 \cdots 0}_{h-b-2} 10 \underbrace{11 \cdots 1}_{b-1} 0.$$

We now follow the necklace in which this state is contained until we arrive at the state

$$1 \underbrace{00 \cdots 0}_{h-b-1} 10 \underbrace{11 \cdots 1}_{b-2},$$

juggle an $(h - b + 1)$-throw to arrive at the state

$$\underbrace{00\cdots0}_{h-b-1}\underbrace{11\cdots1}_{b}0,$$

and follow the necklace in which this state is contained back to the ground state. In doing all this, we have traversed a prime loop of length $2(h - 1)$. This value is greater than h. This shows that a maximal prime loop is nontrivial. It also suggests the following "necklace-hopping" strategy for constructing long prime loops. Start by following one necklace as long as possible, hop to another necklace and follow that one for some time before you hop to the next, and so forth until finally, you hop back onto the first necklace at some state that you have not yet visited to then complete your journey on this necklace. Ideally, on your journey you will visit every necklace and traverse large parts of every one of them. We will see that if you follow any of the known maximal prime loops, you will automatically be doing some serious necklace hopping.

It is possible to use our states-from-throws result (see Subsection 2.8.3) to conclude that a prime juggling sequence whose number of vertices is not too far away from the total number of vertices of the state graph will contain a large number of 0s and hs. Consequently, such a prime loop is really spliced together from large parts of necklaces. To illustrate the reasoning involved in coming to this conclusion, let's assume that there was a prime loop that contains all vertices of the state graph. To calculate exactly how many throws to each of the possible heights the prime loop would contain, we need to know the total number d_i of 1s in the ith positions of the states in the loop, for $i = 1, 2, \ldots, h$, and the total number d_0 of all 0s in the first positions of the states in the loop. In the special case under consideration, these numbers are clearly as follows:

$$d_0 = \binom{h-1}{b} \text{ and } d_i = \binom{h-1}{b-1},$$

for $i = 1, 2, \ldots, h$. This means that the juggling sequence contains d_0 0-throws, d_h h-throws, and $d_i - d_{i+1} = 0$ i-throws for $i = 1, 2, \ldots, h - 1$; that is, the prime loop contains only 0s and hs. More generally, the more states starting with a 0 and the more states ending with a 1 a juggling sequence visits, the more 0s and hs it contains. This suggests that prime juggling sequences whose number of vertices is not too far away from the total number of vertices of the state graph will contain a large number of 0s and hs.

The considerations above also show that there cannot be any prime loop that contains every state of the state graph because this prime loop would be nontrivial and have all necklaces as subgraphs, which is impossible. It is possible to even show that every nontrivial prime loop is missing at least one state from every necklace in the state graph. Assume that this is not the case and that m is a nontrivial prime loop that contains all states of some necklace. Consider one of the states in the necklace contained in the loop. If it starts with a 0, then the 0-edge connecting this state with the next state in the necklace must also be part of the loop. If it starts with a 1, then the next state in the necklace ends with a 1. Since this next state is also contained in m, the h-edge from the first to the second state must also be contained in the loop. This implies that all edges in the necklace and therefore the necklace itself are contained in m, which is a contradiction because m was supposed to be nontrivial and prime.

Hence, for $2 \leq b \leq h - 2$ and $(b, h) \neq (2, 4)$,

$$MP(b, h) \leq Vert(b, h) - Neck(b, h).$$

This upper bound turns out to be sharp in most known cases but not always.

Table 2.5 was computed by Jack Boyce; see [14]. It lists the maximal prime 2-, 3-, 4-, and 5-ball juggling sequences of small heights up to cyclic shifts. For brevity, n consecutive identical elements with the value v have been abbreviated by v^n. In those cases in which $2 \leq b \leq h - 2$, the three associated values $MP(b, h)$, $Vert(b, h) - Neck(b, h)$, and $Vert(b, h)$ are also listed, separated by forward slashes.

In all cases considered by Boyce, the difference between the first two numbers is less than or equal to 2. In fact, in most cases, the two values coincide. Moreover, in most cases, there is a unique maximal prime juggling sequence up to cyclic shifts. Boyce's original list also contains the very long maximal prime juggling sequences in some cases for which we only list the associated three values.

Note that, whenever $h = 2b$ in this list, there are at least two maximal prime juggling sequences up to cyclic shifts. One of the reasons for this is that in these cases the complement of every maximal prime juggling sequence is another maximal prime juggling sequence; see Subsection 2.8.5. To construct the complement of such a juggling sequence, just reverse the order of its elements and then replace every element v by $h-v$. For example, in the case of three balls and height 6, the maximal prime juggling sequence

$$6^2 20606206^2 050^2$$

is the complement of the maximal prime juggling sequence

$$6^2 160^2 64060642^2.$$

However, it is also possible that a maximal prime juggling sequence is its own complement, as in the case of the two maximal prime 2-ball juggling sequences of height 4.

Following the table, we summarize and extend what we have deduced so far about prime loops and maximal prime juggling sequences.

Maximal Prime Juggling Sequences	
two balls	
3: $3^2 0$	
4: 4130	$4/4/6$
$4^2 0^2$	
5: 5205040^2	$8/8/10$
6: $62050^2 6050^3$	$12/12/15$
7: $730^2 70^2 60^3 7060^4$	$18/18/21$
8: $830^2 70^3 80^2 70^4 8070^5$	$24/24/28$
we give element h of this sequence of sequences on page 57.	
three balls	
4: $4^3 0$	
5: $5^2 150530$	$8/\ 8/10$
6: $6^2 20606206^2 050^2$	$15/16/20$
$6^2 206061^2 6^2 050^2$	
$6^2 160^2 64060640^2$	
$6^2 160^2 640606130$	
$6^2 160^2 5^2 060640^2$	
$6350606206^2 050^2$	
7: $7^2 30^2 707170^2 706070^2 740^2 7^2 060^3$	$30/30/35$
8: $8^2 2080^3 850^2 808070^3 8082080^2 8070^2 80^2 850^3 8^2 070^4$	$49/49/56$
$8^2 2080^3 860^2 80^2 8208080^2 70^2 808070^3 80840^3 8^2 070^4$	
$8^2 180^4 86080^3 860^2 80^2 8208080^2 70^2 808070^3 80860^4$	
9: $9^2 50^4 909190^4 908090^4 970^2 90^3 9190^2 90^3 8080^2 90^2 9080^3 90^2 960^4 9^2 080^5$	$74/74/84$
heights 10–17 yield one (very long) maximal sequence each.	
10:108/108/120, 11:149/150/165, 12:200/201/220, 13:263/264/286	
14:337/338/364, 15:424/424/455, 16:524/525/560, 17:639/640/680	
four balls	
5: $5^4 0$	
6: $6^3 1606^2 1640$	$12/12/15$
7: $7^3 170^2 7^2 307^2 071707^2 060707^2 40^2$	$30/30/35$
8: $8^3 30^2 808^2 180^2 80818^2 0^2 8070780^2 8^2 07080^2 8^2 40^2 8^2 0830^2 8^3 070^3$	$58/60/70$
$8^3 180^3 8^2 5080^2 8^2 40^2 8^2 08180^2 8^2 0181808^2 0^2 70808^2 070^2 808^2 50^3$	
heights 9, 10, and 11 one sequence each.	
9: 112/112/126, 10:188/188/210, 11:300/300/330	
five balls	
6: $6^5 0$	
7: $7^4 1707^3 17^2 07^2 40$	$18/18/21$
8: $8^4 20808^3 180808^2 18^2 0808608^2 08^2 208^3 08208^4 070^2$	$49/49/56$
$8^4 180^2 8^3 40808^3 180808^2 18^2 0808608^2 08^2 208^3 0860^2$	
$8^4 180^2 8^3 308^2 08^2 1808^2 0860808^3 180808^2 308^3 0860^2$	
9: five sequences with	$112/112/126$

TABLE 2.5. The maximal prime 2-, 3-, 4-, and 5-ball juggling sequences of small heights up to cyclic shifts and $MP(b,h)/Vert(b,h) - Neck(b,h)/Vert(b,h)$.

Maximal Prime Loops and
Maximal Prime Juggling Sequences

Let $Vert(b, h)$, $Neck(b, h)$, and $MP(b, h)$ be the number of vertices, the number of necklaces, and the length of a maximal prime loop of the b-ball state graph of height h, where $0 \leq b \leq h$ and $h \geq 1$. Then, these three numbers are equal to the corresponding three numbers associated with the $(h - b)$-ball state graph of height h. Furthermore, the number of (maximal) prime loops in the b-ball state graph of height h equals the number of (maximal) prime loops in the $(h - b)$-ball state graph of height h. In particular,

$$MP(0, h) = MP(h, h) = 1 \text{ and } MP(1, h) = MP(h - 1, h) = h,$$

and there is exactly one corresponding maximal prime loop each. The 2-ball state graph of height 4 contains exactly one trivial maximal prime loop and one nontrivial maximal prime loop. The length of these maximal prime loops is

$$MP(2, 4) = 4.$$

For $2 \leq b \leq h - 2$ and $(b, h) \neq (2, 4)$, we have

$$2(h - 1) \leq MP(b, h) \leq Vert(b, h) - Neck(b, h),$$

$$Vert(b, h) - Neck(b, h) = \binom{h}{b} - \frac{1}{h} \sum_{d | gcd(b,h)} \phi(d) \binom{h/d}{b/d}.$$

For $h \geq 4$,

$$MP(2, h) = Vert(2, h) - Neck(2, h) = \begin{cases} \frac{h(h-2)}{2} & \text{if } h \text{ is even,} \\ \frac{(h-1)^2}{2} & \text{if } h \text{ is odd.} \end{cases}$$

Moreover, using the same abbreviating notation as in Table 2.5,

$$he0^{e-1}g0^eh0^{e-1}g0^{e+1}h0^{e-2}g0^{e+2}h0^{e-3}g0^{e+3}h0^{e-4}\cdots h0g0^{h-3},$$

for $h \geq 6$ even, $g = h - 1$, $e = (h - 2)/2$, and

$$he0^{e-1}h0^{e-1}g0^eh0^{e-2}g0^{e+1}h0^{e-3}g0^{e+2}h0^{e-4}g0^{e+3}\cdots h0g0^{h-3},$$

for $h \geq 5$ odd, $g = h - 1$, $e = (h - 1)/2$, are maximal prime 2-ball juggling sequences of height h.

The assertions linking the b-ball state graph of height h and the $(h - b)$-ball state graph of height h follow immediately from the fact that the first graph is the complement of the second graph; see Subsection 2.8.5. The lower bound $2(h - 1)$ for $MP(b, h)$ is the length of the prime loop that we constructed above. In the special case $(b, h) = (3, 5)$, the length of the maximal prime loop equals this lower bound. The maximal prime 2-ball juggling sequences in Table 2.5 suggest that the sequence above for general $h > 4$ should be the only maximal prime 2-ball juggling sequence of height h up to cyclic shifts. It is easy to check that the period of this sequence is the maximum possible one. We leave it to the interested reader to verify that this sequence is really a prime juggling sequence.

2.8.5   Complements of State Graphs

In this subsection, we identify three ways of transforming state graphs into parts of other state graphs.

The *k-extension*, $k \in \mathbf{N}$, of the b-ball state graph G of height h is constructed from G by attaching k 0s to every one of its vertices. Everything else is left unchanged. Clearly, the k-extension is a subgraph of the b-ball state graph of height $h + k$.

The *vertical k-shift*, $k \in \mathbf{N}$, of G is constructed from G by prefixing k 1s in front of every one of the vertices and adding k to every one of the labels of the edges. As an immediate consequence of our result on vertical shifts of juggling sequences (see page 23), we find that this vertical k-shift is a subgraph of the $(b + k)$-ball state graph of height $h + k$.

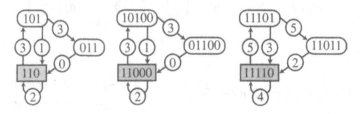

FIGURE 2.24. The 2-ball state graph of height 3 together with its 2-extension and vertical 2-shift.

The 0-ball state graph of height h and the h-ball state graph of height h have only one vertex and one edge each and are clearly the simplest among the state graphs of height h.

The 1-ball state graph of height h and the $(h - 1)$-ball state graph of height h also have very simple structures; see Figure 2.25 for the two graphs in the case $h = 4$.

Note that these two graphs have the same number of vertices and edges and that their underlying undirected and unlabeled graphs coincide. This

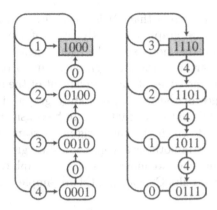

FIGURE 2.25. The 1-ball state graph of height 4 and the 3-ball state graph of height 4.

suggests the following definition. The *complement* of the b-ball state graph of height h is constructed by modifying the state graph as follows:

- Reverse the orientations of all edges.

- Given an edge of the state graph, replace its label i by $h - i$.

- Given a vertex of the state graph, reverse the order of its digits and change every 0 into a 1 and every 1 into a 0. For example, according to this rule, 1111001 is to be replaced by 0110000.

This means that the two graphs in Figure 2.25 are complements of each other. Similarly, the 0-ball state graph of height h and the h-ball state graph of height h are complements of each other. More generally, we derive the following neat result, which was first discovered by Jack Boyce; see [13]:

Complements of State Graphs

(C1) The complement of the b-ball state graph of height h is the $(h - b)$-ball state graph of height h.

(C2) The complement of the complement of a state graph is the state graph itself.

(C3) Given a b-ball juggling sequence $s = \{a_k\}_{k=0}^{p-1}$ of height less than or equal to h, the sequence $c = \{h - a_{p-1-k}\}_{k=0}^{p-1}$ is an $(h - b)$-ball juggling sequence of height less than or equal to h.

(C4) The sequence s is ground-state, excited, or prime if and only if c is.

We first convince ourselves that C1 is true. Clearly, the vertices of the complement of the state graph are exactly those of the $(h - b)$-ball state graph of height h. Furthermore, since the complement of the complement of a state graph is certainly the state graph itself, to prove the result it suffices to show that the edges of the complement of the b-ball state graph of height h are edges of the $(h - b)$-ball state graph of height h. There are two essentially different cases that have to be considered depending on whether the vertex at which the edge originates starts with a 0 or a 1. Figure 2.26 shows that in both cases edges of a state graph really turn into edges of a state graph when we move from a state graph to its complement. The remaining three results C2, C3, and C4 are simple consequences of the main result C1.

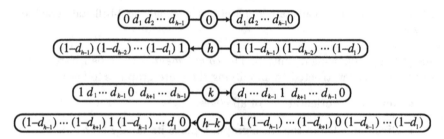

FIGURE 2.26. Corresponding pairs of edges in a state graph of height h and its complement. Note that every one of the d_is is either a 0 or a 1.

By definition, the b-ball state graph of height $2b$ is the complement of itself. This property facilitates a special symmetric drawing of such a graph. Consider the drawings of the 1-ball state graph of height 2 and the 2-ball state graph of height 4 in the left column of Figure 2.27 (S12 and S24) as well as the picture of the 3-ball state graph of height 6 in Figure 2.28. We can complement such a graph—that is, transform it into itself—by first replacing all 0s and 1s of the vertices by 1s and 0s, respectively. Then, we replace all labels of the edges by their "complementary" labels, reverse the orientation of the edges, and, finally, reflect the graph through its "vertical symmetry axis." The last step takes care of reversing the order of the digits in the vertices.

Figures 2.27 and 2.28 illustrate how to build up such a symmetric drawing of the b-ball state graph of height $2b$ from the 1-ball state graph of height 1. As intermediate steps, we come across the c-ball state graph of height $2c$ and the c-ball state graphs of height $2c - 1$ for all $c < b$. Here is how this is accomplished. Suppose we have already constructed the c-ball state graph of height $2c - 1$. Its 1-extension and the complement of this 1-extension are both subgraphs of the c-ball state graph of height $2c$ whose overlap is the 1-extension of the vertical 1-shift of the $(c - 1)$-ball state graph of height $2(c - 1)$. We add all missing vertices and edges to the union of the 1-extension and its complement in a symmetric manner to arrive at a symmetric drawing of the c-ball state graph of height $2c$. Now, the 1-shift of this new graph is a subgraph of the $(c+1)$-ball state graph of height $2(c+1)-1$ whose vertices are exactly those of the full graph whose labels have a leading 1. We draw in the missing vertices and edges to arrive at the complete graph. We summarize what we just described into one step that takes us from a diagram of the c-ball state graph of height $2c-1$ to a diagram of the $(c+1)$-ball state graph of height $2(c+1)-1$ via a symmetric drawing of the c-ball state graph of height $2c$.

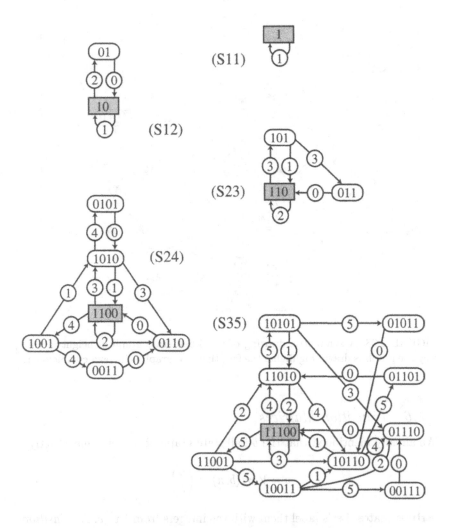

FIGURE 2.27. Successively building up the b-ball state graph of height $2b$ from the 1-ball state graph of height 1 using 1-extensions, vertical 1-shifts, and complements (continued in Figure 2.28).

Finally, note that, as a consequence of Result C4 above, finding one
(maximal) prime loop in the b-ball state graph of height $2b$ means that you
get a second one for free, unless your prime loop is its own "complement";
see also the remarks on page 55.

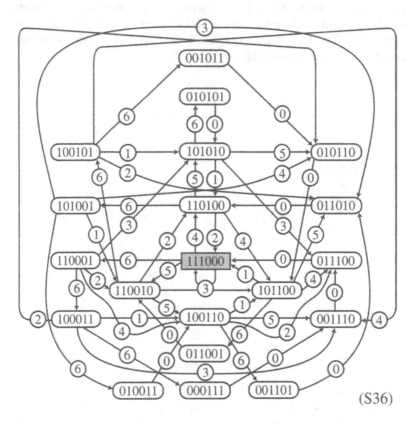

(S36)

FIGURE 2.28. A symmetric drawing of the 3-ball state graph of height 6. The
way the graph is drawn highlights the fact that this graph is its own complement.

2.8.6   Transition Matrices

We already mentioned that the b-ball state graph of height h has exactly

$$v = Vert(b, h) = \binom{h}{b}$$

vertices/states. Let's label them with the integers from 1 to v. A *transition
matrix* of the state graph is the $v \times v$ matrix whose (i, j)th entry is the
integer k if and only if there is an edge in the state graph that originates at
state number i and points at state number j. Figure 2.29 shows a transition

matrix of the 3-ball state graph of height 5 depicted in Figure 2.20. Note that there is only one entry on the diagonal. It corresponds to the only self-edge (labeled b) in the state graph.

	11100	11010	10110	01110	11001	10101	01101	10011	01011	00111
00111								5		▓
01011							5		▓	
10011						5		▓		
01101				5			▓			
10101		5				▓				
11001	5				▓					
01110			4	▓		3		2		0
10110		4	▓		3		1	0		
11010	4	▓		2	1	0				
11100	3	2	1	0						

FIGURE 2.29. A transition matrix corresponding to the 3-ball state graph of height 5.

Of course, it is possible to construct juggling sequences directly in the transition matrix. After some obvious streamlining, constructing a juggling sequence boils down to drawing a special type of closed polygonal path in the matrix whose vertices are either integer entries of the matrix or some of the gray squares marking the main diagonal and whose edges are parallel to either the x-axis or the y-axis. Start with any integer entry in the matrix, move on the horizontal through this entry until you hit the main diagonal, stop, move on the vertical through the point you stopped at to another integer entry, stop, move on the horizontal through this entry until you hit the main diagonal, and so forth. You may end your journey whenever you come across the entry where you started. The first element of the corresponding juggling sequence is the initial entry. The other elements are the entries that you come across on your journey at which you change directions, in the order that you encounter them. A special rule applies to the only integer entry on the diagonal. Whenever you come across it, you can include it in your sequence as many times as you want. In particular, you don't have to include it at all and may treat it as just another gray square. For example, consider the path in Figure 2.30. Here, we start with the 5 the hand points at. Since the path includes the 3 on the main diagonal, it corresponds to the maximal prime 3-ball juggling sequence of height 5,

55150530.

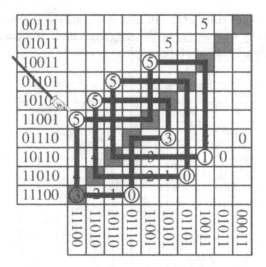

FIGURE 2.30. The closed path in the transition matrix corresponding to the 3-ball juggling sequence 55150530.

On the other hand, it also corresponds to any of the sequences

$$551505033\cdots3.$$

A path constructed as above corresponds to a ground-state juggling sequence if and only if the x-coordinate of the integer entry where you start is the label of the ground state. Finally, we note that all the operations that transform state graphs into (subgraphs of) other state graphs translate in a straightforward manner into the transition-matrix setting. Take, for example, the complement construction outlined in Subsection 2.8.5. Given a transition matrix of the b-ball state graph of height h, replace every integer entry i by $h - i$ and reflect the matrix across its main diagonal to arrive at a transition matrix for the $(h - b)$-ball state graph of height h. If you are also interested in getting the labeling in terms of the different states right as in the figures, then you also have to replace all 0s and 1s in the states by 1s and 0s, respectively.

Further juggling transition matrices are listed in [94], pages 176–179.

3
Multiplex Juggling

All the results derived so far relate to simple juggling patterns and their associated juggling sequences. Simple juggling patterns satisfy Conditions J1–J3; see page 8. In particular, Condition J3 ensures that at most one ball gets caught and thrown on every beat and, if one is caught, the same ball is thrown. In this chapter, we investigate *multiplex juggling patterns* and *multiplex juggling sequences*, which are natural generalizations of the simple juggling patterns and simple juggling sequences. A juggling pattern is a multiplex juggling pattern if it satisfies Conditions J1 and J2 and, in addition, all the balls that get caught on a beat also get tossed on the same beat. This definition implies that every simple juggling pattern is also a multiplex juggling pattern.

To be able to easily distinguish between objects relating to simple juggling sequences and their generalizations, we will add a "simple" to their names. For example, we will speak about simple juggling sequences and simple state graphs instead of just juggling sequences and state graphs.

Most of the notations, conventions, results, and proofs relating to simple juggling sequences have natural counterparts for multiplex juggling sequences. In the following, we will sketch these counterparts. We take the opportunity to present some of these generalizations in ways that deviate from those followed previously and also shed new light on the results that are being generalized.

Multiplex juggling sequences are finite sequences consisting of nonempty (ordered) sets of nonnegative integers that record what kinds of throws are made on every beat. Furthermore, if one of the sets contains more than one element, then all the elements in the set are positive integers. As a first example, let us consider the multiplex juggling sequence

$$\{1,4\}\{1\}.$$

This juggling sequence has period 2. On the first beat, two balls get caught. On the same beat, one of these balls is thrown to height 1 and the other to height 4. On the next beat, one ball is caught and thrown as a 1-throw. In the juggling literature, this multiplex juggling sequence is usually written in the form

$$[14]1;$$

that is, curly brackets are replaced by square brackets and one-element sets by the elements contained in them. Furthermore, commas are omitted as in the case of simple juggling sequences whenever it is understood that we are only dealing with throws up to height 9; see our respective convention on page 9.

As in the case of simple juggling patterns and sequences, there are lots of different multiplex juggling sequences that describe a given multiplex juggling pattern. We arrive at these different sequences from a minimal juggling sequence such as [14]1 by cyclic shifts, forming multiple copies, as well as (and this is new) permutations within square brackets. For example, the multiplex juggling sequences

$$1[14], [14]1[14]1[14]1, [41]1, \text{ and } 1[14]1[41]$$

all describe the same multiplex juggling pattern.

Figure 3.1 shows the juggling diagram of [14]1.

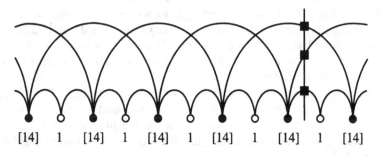

FIGURE 3.1. The juggling diagram of the multiplex juggling sequence [14]1.

3.1   Average Theorem and Permutation Test

How many balls are necessary to juggle [14]1? The juggling diagram is not the union of disjoint orbits as in the case of a simple juggling sequence, so we cannot count orbits to establish how many balls are being juggled. However, at any point in time that does not correspond to a beat, all the balls are in the air. This means that the number of intersections of any vertical line with the arcs, which is not passing through one of the points of intersection of two arcs, is equal to the number of balls juggled. In Figure 3.1, we have drawn in one such line and the points of intersection with the arcs. It turns out that [14]1 is a 3-ball multiplex juggling sequence. The Average Theorem for simple juggling sequences (see page 15) generalizes as follows:

The Average Theorem for Multiplex Juggling Sequences

The number of balls necessary to juggle a multiplex juggling sequence equals the sum of the integers in the sequence divided by its period.

To prove this, consider a multiplex juggling sequence s of period p. Furthermore, let $sum(s)$ and $balls(s)$ denote the sum of the integers in s and the number of balls required to juggle s, respectively. Let I_n, with $n \in \mathbf{N}$, be the time interval starting at beat 0 and ending on beat np. During each of the np unit time intervals in I_n, starting on beats, we look at the number of balls in the air and form the sum of these numbers. It is clear that this sum is $np\,balls(s)$. On the other hand, for large n, it is not hard to see that this sum is also equal to $n\,sum(s)$. We divide both expressions by np to arrive at the desired result

$$\frac{sum(s)}{p} = balls(s).$$

The permutation test for simple juggling sequences also generalizes.

The Permutation Test for Multiplex Juggling Sequences

Let $s = \{S_i\}_{i=0}^{p-1}$ be a sequence of nonempty ordered sets of nonnegative integers such that none of the sets S_i with more than one element contains any 0s. Furthermore, let $s' = \{S_i'\}_{i=0}^{p-1}$, where $S_i' = (i + S_i) \bmod p$. Then, s is a multiplex juggling sequence if and only if the total number of is in s' equals the number of elements of the ith element of s' for $i = 0, 1, \ldots, p - 1$.

For example, if

$$s = \{2\}, \{2\}, \{7, 2\}, \{5, 4\}, \{2, 1\}, \text{ (that is, } s = 22[72][54][21]),$$

then $p = 5$ and

$$
\begin{aligned}
s' &= \{0+2\}, \{1+2\}, \{2+7, 2+2\}, \{3+5, 3+4\}, \{4+2, 4+1\} \bmod 5 \\
&= \{2\}, \{3\}, \{4, 4\}, \{3, 2\}, \{1, 0\}.
\end{aligned}
$$

There are a total of one 0, one 1, two 2s, two 3s, and two 4s in s'. Hence, we are dealing with a multiplex juggling sequence.

We only sketch a proof of the result above. Let $s = \{S_i\}_{i=0}^{p-1}$ be one of the sequences under discussion. As usual, we perform all the throws in S_i on all beats that are equal $i \bmod p$. Then, s is a multiplex juggling sequence if and only if for all $i \in \mathbf{Z}$ the number of balls caught on beat i equals the number of nonzero elements in S_i. Now, it is easy to see that a k-throw, $k \neq 0$, in one of the sets S_i contributes one ball to the balls landing on beat j if and only if $(k + i) \bmod p = j \bmod p$. Note that 0-throws take care of themselves. A straightforward rephrasing of the last two statements yields the result we are after.

3.2   Number of Multiplex Juggling Sequences

We calculate the number of b-ball multiplex juggling sequences of period p using an extension of the juggling cards idea introduced in Section 2.7.

For every k, with $0 \leq k \leq b$, we construct a deck $D_{b,k}$ consisting of infinitely many cards. There are exactly $\binom{b}{k}$ different cards contained in every deck, and every one of these cards is represented infinitely many times in the deck. The construction principle for all decks is illustrated in Figure 3.2, which shows the $6 = \binom{4}{2}$ different cards in the deck $D_{4,2}$. These cards capture all possible ways to choose k balls out of the b balls in the air, have the juggler catch these k balls, and, finally, throw these k balls up so that they are placed in the lowest orbits.

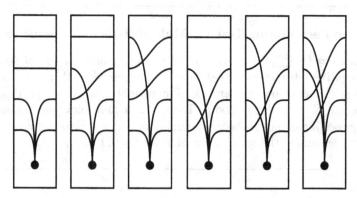

FIGURE 3.2. The $6 = \binom{4}{2}$ juggling cards for multiplex juggling at most four balls, where two balls are caught and thrown on every beat.

We are interested in constructing multiplex juggling sequences with at most b balls using these decks of cards. Let t_i, $i = 0, 1, \ldots, p - 1$, be nonnegative integers less than or equal to b. We draw a card from deck D_{b,t_0} and put it on the table, draw a card from deck D_{b,t_1} and put it next to the first card on the right, draw a card from deck D_{b,t_2} and put it next to the second one on the right, and so forth until there are p cards on the table. We repeat this arrangement of cards infinitely often to the left and right to arrive at a juggling diagram of a multiplex juggling sequence of period p with at most b balls in which t_i balls are caught and thrown on beat $i = 0, 1, \ldots, p - 1$, respectively. It is clear how to read off from the diagram which throws are made on beat i. If there is more than one throw being performed on this beat, we note these throws in the order that the corresponding arcs emanate from the beat point. Figure 3.3 shows juggling diagrams for the multiplex juggling sequences 21[51] and 21[15] constructed with multiplex juggling cards.

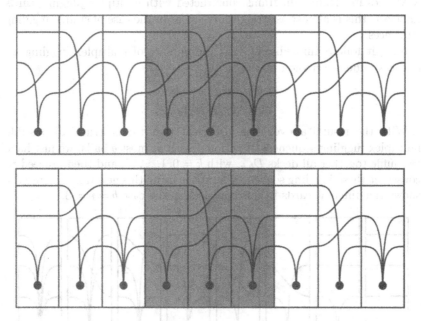

FIGURE 3.3. The multiplex juggling sequences 21[51] and 21[15] constructed with multiplex juggling cards.

Note that in the upper diagram the 5-throw in the multiplex [51]-throw really starts below the 1-throw, and in the lower diagram the 5-throw starts above the 1-throw. Also, the two corresponding arcs in the upper diagram intersect in exactly one point that is not a beat point, whereas in the lower diagram the two corresponding arcs do not intersect. This is true in general: Given two throws a and b in a multiplex juggling sequence that are performed on a certain beat such that a precedes b within the pair of square brackets $[\cdots a \cdots b \cdots]$, then the two corresponding arcs will not intersect in a nonbeat point if and only if $a \le b$ and will intersect in exactly one such point otherwise. On the other hand, arcs that start at different beat points will intersect in exactly one nonbeat point, one beat point, or no point if and only if the end point of the arc that starts first comes before, is the same, or comes after the end point of the arc that starts later, respectively. Finally, we note that every point of intersection of two arcs occurs as far to the right as possible.

It turns out that every multiplex juggling sequence under discussion can be constructed exactly once using the sets of multiplex juggling cards described above. The proof for this fact is a simple extension of the one given in the case of simple juggling sequences; see [36] for details. In particular, it involves devising an algorithm that, on input of a multiplex juggling sequence, outputs the representation of the corresponding juggling diagram in terms of juggling cards. Since we have complete information about the ways arcs in juggling diagrams constructed with multiplex juggling cards intersect, this is almost as straightforward as in the case of simple juggling sequences.

Now, it follows immediately that the number of multiplex juggling sequences under discussion is

$$\binom{b}{t_0}\binom{b}{t_1}\cdots\binom{b}{t_{p-1}}.$$

With the results above, it is also clear that to construct all possible multiplex juggling sequences of period p with at most b balls, we just have to shuffle together all decks $D_{b,k}$, with $k = 0, 1, \ldots, b$, and then proceed to construct these juggling sequences as above using this new deck. Figure 3.4 shows the different cards in this new deck in the case $b = p = 3$.

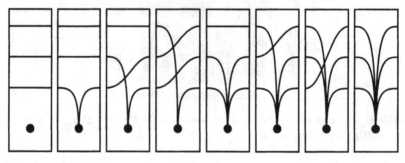

FIGURE 3.4. The $\binom{3}{0} + \binom{3}{1} + \binom{3}{2} + \binom{3}{3} = 2^3 = 8$ multiplex juggling cards used for constructing all multiplex juggling sequences with at most three balls.

Hence, the total number of such juggling sequences is

$$\left(\sum_{i=0}^{b}\binom{b}{i}\right)^p = (2^b)^p = 2^{pb}.$$

Numbers of Multiplex Juggling Sequences

(M1) The number of all multiplex juggling sequences of period p and at most b balls is

$$M^{\leq}(b,p) = 2^{pb}.$$

(M2) The number of all b-ball multiplex juggling sequences of period p, with $b \geq 1$, is

$$M(b,p) = M^{\leq}(b,p) - M^{\leq}(b-1,p) = 2^{pb} - 2^{p(b-1)}.$$

For example, for $b = p = 3$, we calculate $M(3,3) = 2^9 - 2^6 = 448$. However, up to cyclic shifts and permutations inside square brackets, as detailed above, there are only 66 essentially different 3-ball multiplex juggling sequences of period 3. These patterns are listed in Table 3.1.

333 (1/0)	[21]24 (6/2)	[33]03 (3/1)	[111][111][111] (1/0)
423 (3/1)	[21]4[11] (6/2)	[33]12 (3/1)	[211][11][111] (9/3)
441 (3/1)	[21]51 (6/2)	[33]30 (3/1)	[211][21][11] (18/6)
450 (3/1)	[21][21][21] (8/2)	[41][11][11] (6/2)	[211][22]1 (9/3)
522 (3/1)	[22]2[21] (6/2)	[41][21]1 (12/4)	[221]1[111] (9/3)
531 (3/1)	[22]3[11] (3/1)	[41][22]0 (6/2)	[221]2[11] (9/3)
603 (3/1)	[31][11][21] (12/4)	[42]1[11] (6/2)	[222]0[111] (3/1)
612 (3/1)	[31][21]2 (12/4)	[42]21 (6/2)	[311][11][11] (9/3)
630 (3/1)	[31]14 (6/2)	[43]11 (6/2)	[311][21]1 (18/6)
711 (3/1)	[31]23 (6/2)	[43]20 (6/2)	[311][22]0 (9/3)
720 (3/1)	[31]41 (6/2)	[51]1[11] (6/2)	[321]1[11] (18/6)
801 (3/1)	[31]50 (6/2)	[51]21 (6/2)	[321]21 (18/6)
900 (3/1)	[31][31]1 (12/4)	[52]0[11] (6/2)	[331]11 (9/3)
	[32]04 (6/2)	[53]01 (6/2)	[331]20 (9/3)
	[32]0[31] (12/4)	[61]11 (6/2)	[332]01 (9/3)
	[32]1[21] (12/4)	[61]20 (6/2)	[322]0[11] (9/3)
	[32]22 (6/2)	[62]01 (6/2)	[333]00 (3/1)
	[32]31 (6/2)	[63]00 (6/2)	

TABLE 3.1. The 66 essentially different 3-ball multiplex juggling sequences of period 3.

Also listed behind every sequence are two numbers separated by a forward slash. The first of these is the number of different multiplex juggling sequences that this particular sequence represents. You can check for yourself that these numbers indeed add up to 448.

Note that the multiplex juggling sequences on the left are the 13 essentially different 3-ball simple juggling sequences of period 3. We could still get rid of the juggling sequences that are not minimal (that is, the sequences 333 and [111][111][111]) and the two nonminimal juggling sequences represented by [21][21][21] (that is, [21][21][21] and [12][12][12]). It would be nice to have a simple formula for the number of essentially different minimal b-ball juggling sequences of period p. Such a formula would be the multiplex counterpart of the formula for $MS(b, p)$, the number of essentially different minimal b-ball simple juggling sequences of period p; see page 40. Using the same argument as we used there, the number of all minimal b-ball multiplex juggling sequences of period p up to cyclic shifts, with $b \geq 1$, can be calculated to be

$$\frac{1}{p} \sum_{d|p} \mu\left(\frac{p}{d}\right)\left(2^{db} - 2^{d(b-1)}\right).$$

Of course, this is not quite what we want since it does not get rid of duplications that arise from permutations in square brackets. However, if we are interested in an upper bound for the number we are really interested in, this one is already much better than $2^{pb} - 2^{p(b-1)}$. In the case $b = p = 3$, this formula tells us that, up to cyclic shifts, there are 148 minimal 3-ball multiplex juggling sequences of period 3. The second number behind one of the multiplex juggling sequences in Table 3.1 is the number this entry contributes to 148.

By shuffling decks $D_{b,0}$ and $D_{b,1}$ together, we arrive at the deck that we used to construct all simple juggling sequences with at most b balls. In general, it is possible to generate and count a host of different classes of multiplex juggling sequences by combining the different decks of cards $D_{b,i}$. For example, from a practical point of view, the b-ball multiplex juggling sequences in which at most two balls are caught and thrown on every beat are probably the most interesting ones. We arrive at the corresponding deck of cards by shuffling together decks $D_{b,0}$, $D_{b,1}$, and $D_{b,2}$ and calculate the corresponding number of b-ball multiplex juggling sequences of period p, with $b \geq 2$, to be

$$\left(1 + b + \binom{b}{2}\right)^p - \left(1 + (b-1) + \binom{b-1}{2}\right)^p$$

$$= \left(\frac{b^2 + b + 2}{2}\right)^p - \left(\frac{b^2 - b + 2}{2}\right)^p.$$

The number of such sequences up to cyclic shifts that are minimal is

$$\frac{1}{p} \sum_{d|p} \mu\left(\frac{p}{d}\right)\left(\left(\frac{b^2 + b + 2}{2}\right)^d - \left(\frac{b^2 - b + 2}{2}\right)^d\right).$$

3.3   Weights of Multiplex Juggling Sequences

Given a multiplex juggling sequence s of period p, construct its juggling diagram in terms of multiplex juggling cards and focus on all the points of intersection of arcs on p cards that form a period. Let $cross(s)$ be the number of those points of intersection that are not beat points. For example, $cross(12[51]) = 3$ and $cross(12[15]) = 2$; see Figure 3.3. Furthermore, let the *weight* of s be the formal expression $q^{cross(s)}$. We have already seen in the case of simple juggling sequences that these terms have very interesting and useful combinatorial interpretations. Now, it is a straightforward exercise to calculate the sums of weights of the multiplex juggling sequences of period p with at most b balls; see the very nice paper by Richard Ehrenborg and Margaret Readdy [36] for details about this and the following results.

The Sum of the Weights

The sum of the weights of all multiplex juggling sequences of period p with at most b balls, such that t_i, $i = 0, 1, \ldots, p - 1$ is the number of balls thrown on beat i, is the product of *Gaussian coefficients*[1]

$$\prod_{i=0}^{p-1} \begin{bmatrix} b \\ t_i \end{bmatrix}_q.$$

The sum of the weights of all multiplex juggling sequences of period p with at most b balls is the product

$$\left(\sum_{i=0}^{b} \begin{bmatrix} b \\ i \end{bmatrix}_q \right)^p = G_b(q)^p.$$

Here, $G_b(q)$ denotes a *Galois number*.[1]

A large number of important combinatorial results involve Gaussian coefficients and Galois numbers. It was demonstrated in [36] how some of these results can be proved, very elegantly and naturally, completely within the framework of juggling sequences. We already mentioned the simple ex-

---

[1] *Gaussian coefficients and Galois numbers.* The Gaussian coefficients are defined for all pairs of nonnegative integers k and n such that $0 \leq k \leq n$:

$$\begin{bmatrix} n \\ 0 \end{bmatrix}_q = \begin{bmatrix} n \\ n \end{bmatrix}_q = 1, \begin{bmatrix} n \\ k \end{bmatrix}_q = \frac{[n]_q!}{[k]_q! \cdot [n-k]_q!}, 1 \leq k \leq n - 1, \text{ with } [l]_q = \sum_{i=0}^{l-1} q^i, [l]_q! = \prod_{i=1}^{l} [i]_q.$$

pression for the Poincaré series of the Weyl group $\tilde{A}_{p-1}$ in terms of weights. Further examples include the following identity, which was originally proved by Carlitz in [20]:

$$[b]_q{}^p = \sum_{i=0}^{p} S_q[p, i] \cdot [i]_q! \cdot \begin{bmatrix} b \\ i \end{bmatrix}_q,$$

where $S_q[p, i]$ denotes a *q-Stirling number of the second kind*.[2]

In the proof of this identity, the sum of the weights of the simple juggling sequences of period p with at most b balls and no 0-throws is calculated and expressed in two different ways corresponding to the left-hand and right-hand sides of this identity. The left-hand side is just a concise way of writing this sum as we calculated and expressed it in Section 2.7. The second way is the result of ingenious counting based on the following interesting observations and results:

- Every simple juggling sequence s under consideration induces a partition into blocks of the set consisting of the beat points $0, 1, \ldots, p-1$. The individual blocks are the intersections of the orbits of the juggling diagram of s with this set of beat points.

- The intertwining number of this partition equals the number of crossings in the juggling diagram of s "between" beat points 0 and $p-1$.

- The identity
$$S_q[p, i] = \sum_{\pi} q^{int(\pi)}, p, i \geq 1,$$

  where the sum is over all partitions π of the set $\{0, 1, \ldots, p-1\}$ into i blocks and $int(\pi)$ denotes the intertwining number of π.

The Galois number $G_n(q)$ is defined in terms of Gaussian coefficients:

$$G_n(q) = \sum_{i=0}^{n} \begin{bmatrix} n \\ i \end{bmatrix}_q.$$

If we set $q = p^r$, where p is a prime, then the Gaussian coefficient above counts the subspaces of dimension k of an n-dimensional vector space over the finite field with q elements, and the Galois number counts all subspaces of this vector space. If we let q tend to 1, the Gaussian coefficient tends to the binomial coefficient $\binom{n}{k}$ and the Galois number to 2^n. In fact, the Gaussian coefficients are q-analogues of the binomial coefficients and share many of their properties; see [102], Subsection 3.4.1, for more detailed information about Gaussian coefficients and Galois numbers.

[2] *Stirling numbers and q-Stirling numbers of the second kind.* The q-Stirling numbers of the second kind are defined by the recursion formula

$$S_q[n, k] = q^{k-1} \cdot S_q[n-1, k-1] + [k]_q \cdot S_q[n-1, k],$$

where $n, k \geq 1$. Define $S_q[n, k] = \delta_{n,k}$ for $n = 0$ or $k = 0$. If we let q tend to 1, then $S_q[n, k]$ tends to $S(n, k)$, one of the Stirling numbers of the second kind, which counts the number of partitions of the set $\{1, 2, \ldots, n\}$ into k blocks.

- By "contracting" the beat points $kp, kp + 1, \ldots, kp + p - 1$ in the juggling diagram of s into a new vertex k for all integers k and discarding all arcs whose end points get identified under contraction, we arrive at a juggling diagram of a multiplex juggling sequence of period 1 with the same number of balls as s. If $s = \{a_i\}_{i=0}^{p-1}$, omit all 0s from the sequence $\{b_i\}_{i=0}^{p-1}$, where b_i is the integer part of $(a_i + i)/p$, and enclose what is left in square brackets to arrive at the contracted multiplex juggling sequence. The number of integers in this new juggling sequence equals the number of blocks of the partition associated with s. For example, the 4-ball simple juggling sequence 51419 contracts to the 4-ball multiplex juggling sequence [112], and the partition of the beat points 0, 1, 2, 3, and 4 associated with 51419 consists of the three blocks $\{0\}, \{1, 2\}, \{3, 4\}$.

To come up with the right-hand side of the identity, it now suffices to keep track of what happens to a crossing that counts towards $cross(s)$ under contraction. Four cases need to be considered depending on how the end points of the two arches that intersect in this crossing get identified. One case is accounted for by the factor $[i]_q!$, one by the factor $S_q[p, i]$, and two by the Gaussian coefficient $\left[\begin{smallmatrix} b \\ i \end{smallmatrix}\right]_q$.

We only note that if we let q tend to 1, the identity above turns into

$$b^p = \sum_{i=0}^{p} S(p, i) i! \binom{b}{i},$$

where $S(p, i)$ is a *Stirling number of the second kind*.

Apart from the applications discussed so far, [36] also includes some results that express q-analogues of the numbers that count unitary decompositions of vectors in terms of Gaussian coefficients. The ideas used to prove these results are very similar to the ones described above.

The original proof in [19] that $(b + 1)^p$ is the number of simple juggling sequences of period p with at most b balls involves other important combinatorial concepts, numbers, and identities, such as the Eulerian numbers, ascents and descents of permutations, and Worpitzky's identity. In view of what has already been achieved in [17], [19], and [36], it seems likely that a detailed review of the results involving Eulerian numbers, Stirling numbers, and Gaussian coefficients will reveal more ways in which juggling sequences can be applied within combinatorics; for some hints as to what to look for, see [36], Section 9.

3.4 Multiplex State Graphs

The b-ball simple state graph of height h, with $0 \le b \le h$ and $h \ge 1$, extends in a natural fashion to the (much larger) *b-ball multiplex state graph of height h.* Figure 3.5 illustrates how this extension is constructed using the example of the 3-ball multiplex state graph of height 3. Note that the simple state graph has the ground state 111 as its only vertex, whereas the multiplex state graph has ten vertices. This particular multiplex state graph has three self-edges starting and pointing at the vertices labeled 111, [111]00, and [11]10. The more 1s that are contained in the first element of a state, the more edges that start at this state. For example, in Figure 3.5, there are 1, 3, 6, and 10 edges starting at any vertex whose first element is a 0, 1, [11], and [111], respectively.

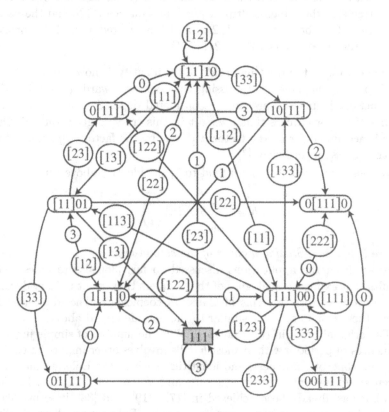

FIGURE 3.5. The 3-ball multiplex state graph of height 3.

Whereas the *b*-ball simple state graph of height *h* could only be defined in a meaningful way for integer choices of *b* and *h* satisfying $0 \le b \le h, h \ge 1$, the *b*-ball multiplex state graph of height *h* can be defined if we choose *b* and *h* such that $b = 0, h \ge 1$ or $b, h \ge 1$. Clearly, the 0-ball and 1-ball multiplex state graphs of height *h* are exactly the 0-ball and 1-ball simple state graphs of height *h*, respectively. For the rest of this section, we will assume that $b, h \ge 1$.

The number of vertices $MVert(b, h)$ of the b-ball multiplex state graph
of height h is the same as the number of ways in which b indistinguishable
objects (the 1s) can be placed into h distinguishable urns (the h positions
in every label). It is a well-known fact that this number is

$$\binom{b + h - 1}{b}.$$

To see this directly, consider a permutation of the b 1s and $h - 1$ indistinguishable separators, say $+$ signs. Then, the $+$ signs separate the 1s into h
groups, and we can interpret the permutation as a b-ball multiplex state of
height h whose ith position is filled with the ith group if it contains 1s and
with a 0 otherwise. For example, $+11+1$ corresponds to the state $0[11]1$.
Conversely, every b-ball multiplex state of height h corresponds to exactly
one such permutation. The number of these permutations is clearly given
by the expression above.

Furthermore, the b-ball multiplex juggling sequences of height at most h
correspond to the directed closed paths in the b-ball multiplex state graph
of height h that start and end at the same state.

3.4.1 Prime Multiplex Juggling Sequences and Loops

We define *(maximal) prime loops* in multiplex state graphs as well as *(maximal) prime multiplex juggling sequences* as in the case of simple juggling.
For example, the 3-ball multiplex juggling sequence

$$23[23]0[33]20[123]$$

is a maximal prime 3-ball multiplex juggling sequence of height at most 3
that corresponds to the circular path highlighted in Figure 3.5, starting
and ending at the ground state. Up to cyclic shifts, there are three more
maximal prime 3-ball multiplex juggling sequences of height at most 3.
These are $23[22]0[331]30[32]$ and $1[33]20[332]03[31]$, both starting at the
ground state, as well as the multiplex juggling sequence $[33]20[332]03[32]0$,
starting at the state $[11]10$.

Remember that a simple state graph of height h has a partition into
simple necklaces; that is, prime loops all of whose edges are labeled either 0
or h. Starting with a state, it is a straightforward exercise to construct the
simple necklace in which it is contained. For example, starting with the
state 11001, we travel along this simple necklace by juggling the juggling
sequence 55005. To construct this juggling sequence, just replace every 1
in the state by an h. Similarly, a multiplex state graph of height h can
be partitioned into *multiplex necklaces*. All edges of a multiplex necklace
are labeled 0, h, $[hh]$, $[hhh]$, and so forth. Again, starting with a state,
we can traverse the multiplex necklace in which it is contained by juggling
the multiplex juggling sequence that results by replacing every 1 in the

state by an h. For example, the juggling sequence that corresponds to the juggling state 01[11] is the multiplex juggling sequence $0h[hh]$. Note that, by definition, every necklace in the b-ball simple state graph of height h is a multiplex necklace in the b-ball multiplex state graph of height h. This means that the partition of the second graph into multiplex necklaces extends the partition of the first graph into necklaces.

The number of multiplex necklaces in the b-ball multiplex state graph of height h is

$$MNeck(b, h) = \frac{1}{b+h} \sum_{d|gcd(b,h)} \phi(d) \binom{(b+h)/d}{b/d},$$

where $gcd(b, h)$ is the greatest common divisor of b and h and the function ϕ is the Euler function; see page 52 for a definition. Furthermore,

$$MNeck(b, h) = MNeck(h, b).$$

Keeping in mind that $gcd(b, h) = gcd(b + h, h)$, this just says that the number of b-ball multiplex necklaces of height h equals the number of b-ball simple necklaces of height $b + h$. To prove this, start with a b-ball multiplex juggling state of height h, prefix it by a $+$ sign and insert $+$ signs in between its h positions, remove all brackets and 0s, and, finally, replace all $h + $ signs by 0s. The end result is a b-ball simple juggling state of height $b+h$ with a leading 0. In fact, it is easy to see that the operation we just described establishes a one-to-one correspondence between the b-ball multiplex juggling states of height h and the b-ball simple juggling states of height $b + h$ with leading 0. Furthermore, this operation and its converse preserve the property of two states to be contained in the same necklace. Finally, it is clear that every simple necklace contains one state with a leading 0. We conclude that the two numbers we are interested in are indeed equal.

We are interested in the length $MM(b, h)$ of maximal prime loops in the b-ball multiplex state graph of height h. We call a prime loop in the multiplex state graph *trivial* if it is one of the necklaces or one of the one-edge loops. There is at least one necklace of length h in the b-ball multiplex state graph of height h containing the state $[11 \cdots 1]00 \cdots 0$. Also, h is the maximum length of a necklace. Furthermore, since the number of vertices in the graph is an absolute upper limit for $MM(b, h)$, it is clear that

$$h \leq MM(b, h) \leq MVert(b, h).$$

In particular, since all states in the 1-ball multiplex state graph of height h are contained in one necklace, we find that

$$MM(1, h) = h.$$

It is also clear that

$$MM(b, 1) = 1.$$

Let $b, h \geq 2$. We first show that in the b-ball multiplex state graph of height h the state

$$S = \underbrace{00 \cdots 0}_{h-1} [\underbrace{11 \cdots 1}_{b}]$$

plays a special role. Assume that a prime loop in this state graph contains S. Then, this loop clearly contains all the states and all the edges labeled 0 of the necklace that contains S. Since there is only one edge in the state graph pointing at S, the prime loop also has to contain this edge and, consequently, the whole necklace that contains S. Hence, the prime loop has to coincide with this necklace, and no nontrivial prime loop can contain S. For example, the highlighted maximal prime loop in Figure 3.5 does not contain S. Since $h < MVert(b, h)$, we can now improve our upper bound above to

$$MM(b, h) \leq MVert(b, h) - 1.$$

For $h = 2$, it is easy to show that

$$[1\underbrace{22 \cdots 2}_{b-1}]2[1\underbrace{22 \cdots 2}_{b-2}][22][1\underbrace{22 \cdots 2}_{b-3}][222] \cdots [\underbrace{22 \cdots 2}_{\frac{b-1}{2}}][\underbrace{11 \cdots 1}_{\frac{b-1}{2}+1}]$$

for b odd and

$$[1\underbrace{22 \cdots 2}_{b-1}]2[1\underbrace{22 \cdots 2}_{b-2}][22][1\underbrace{22 \cdots 2}_{b-3}][222] \cdots [1\underbrace{22 \cdots 2}_{\frac{b}{2}}][\underbrace{11 \cdots 1}_{\frac{b}{2}}]$$

for b even are prime loops of period b. Since $MVert(b, 2) - 1 = b$, these prime loops are maximal and, consequently,

$$MM(b, 2) = b.$$

Now, let's assume that $b \geq 2$ and $h \geq 3$. We improve the lower bound for $MM(b, h)$ by constructing a nontrivial maximal prime loop of period $2h - 1$. For this, we follow the necklace of length h that contains the state

$$\underbrace{00 \cdots 0}_{h-2}1[\underbrace{11 \cdots 1}_{b-1}]$$

to the state

$$[\underbrace{11 \cdots 1}_{b-1}]\underbrace{00 \cdots 0}_{h-2}1$$

and then juggle a $[\underbrace{(h-1)(h-1) \cdots (h-1)}_{b-1}]$-throw to arrive at state

$$\underbrace{00 \cdots 0}_{h-2}[\underbrace{11 \cdots 1}_{b}]0.$$

We now follow the necklace containing this state in until we arrive at the state

$$\underbrace{[11\cdots1]}_{b}\underbrace{00\cdots0}_{h-1}$$

and juggle a $[\overbrace{hh\cdots h}^{b-1}(h-1)]$-throw to arrive back at the state with which we started. Clearly, in doing all this, we have traversed a prime loop of length $2h - 1$. Since this value is greater than h, we can be sure that this prime loop is nontrivial.

On the other hand, it is important to realize that this particular prime loop contains all states of the necklace with which we started. Similarly, the highlighted maximal prime loop in Figure 3.5 contains all states of the necklace that contains the state $[11]10$. Remember that in the case of simple state graphs, we showed that it is impossible for a nontrivial prime loop to contain all states of a necklace. Based on this fact, we were then able to deduce a very good upper bound for $MP(b, h)$ involving the number of necklaces in the graph. Nothing of this sort seems to be possible for multiplex state graphs.

Summarizing, we arrive at the following result:

Maximal Prime Loops

Let $MM(b, h)$ be the number of vertices of a maximal prime loop of the b-ball multiplex state graph of height h, with $b, h \geq 1$. Then,

$$MM(1, h) = h, MM(b, 1) = 1, \text{ and } MM(b, 2) = b.$$

For $b \geq 2$ and $h \geq 3$,

$$2h - 1 \leq MM(b, h) \leq MVert(b, h) - 1 = \binom{b + h - 1}{b} - 1.$$

If the b-ball simple state graph of height h exists—that is, if $h \geq b$—then

$$MP(b, h) \leq MM(b, h).$$

For $b, h \geq 2$, the state

$$\underbrace{00\cdots0}_{h-1}\underbrace{[11\cdots1]}_{b}$$

is not contained in any nontrivial prime loop of the b-ball multiplex state graph of height h.

3.4.2 Throws from States

The following related result can be proved in the same way as the corresponding result in the case of simple juggling sequences; see Subsection 2.8.3:

States Determine Throws

If s and s' are two b-ball multiplex juggling sequences that visit the same states in the b-ball multiplex state graph of some height h the same number of times each, then s contains the same throws the same number of times as s'.

As an example, consider the two 3-ball multiplex juggling sequences of period 3

$$[11][122]2$$
$$[22]1[112]$$

that both start at state $[11]10$ (the uppermost vertex in Figure 3.5) and visit the same three states $[11]10$, $[111]00$, and $1[11]0$.

3.5 Operations Involving Juggling Sequences

In the previous chapter, we encountered a number of different operations that allow us to turn simple juggling sequences into new simple juggling sequences. Most of these operations yield multiplex juggling sequences when applied to multiplex juggling sequences. In the following, we draw up a list of all these operations, complemented by some operations that have not been mentioned before.

We first list the operations that transform simple juggling sequences into simple juggling sequences and multiple juggling sequences into multiple juggling sequences.

- *Multiple copies.* By concatenating multiple copies of a juggling sequence, we get, at least formally, a new juggling sequence. For example, $[41]1$, $[41]1[41]1$, $[41]1[41]1[41]1$, and so forth are all juggling sequences.

- *Cyclic shifts.* By cyclically permuting a juggling sequence, we also arrive at a new juggling sequence.

- *Permutations within square brackets.* Permutation of the entries in one of the square brackets of a multiplex juggling sequence yields a new multiplex juggling sequence.

- *Site swaps.* We defined in Section 2.5 what we mean by transforming a simple juggling sequence using a site swap. This definition generalizes in a straightforward manner to one of site swaps for multiplex juggling sequences. Again, it is easy to show that applying a site swap to a b-ball multiplex juggling sequence yields a new b-ball multiplex juggling sequence. As an example, let's again consider the multiplex juggling sequence $s = 22[72][54][21]$. Then, the 7-throw can be swapped with any of the other throws in the sequence. Here, swapping it with the 2-throw in the set $[72]$ just means that nothing changes (because the distance in terms of positions between the two numbers is 0). Let's swap it with the 2 in the set $[21]$. The distance between the 7 and this 2 is 2. This means that by performing this site swap we construct the sequence

$$22[(2 + 2)2][54][(7 - 2)1] = 22[42][54][51].$$

Sometimes, after a site swap, we end up with a 0 within some square brackets. This 0 can then be deleted. Furthermore, if after the deletion of the 0 only one element is left within the brackets, we can also delete the brackets. Conversely, we can enclose one of the single-throw elements in a juggling sequence by brackets, add a 0 within the brackets, and then perform a site swap that involves this 0 to turn it into a positive integer. For example, start with the simple juggling sequence 441, first turn it into $4[40]1$, and then via a site swap turn it into $1[43]1$.

- *Inverses.* The juggling diagram of the inverse of a juggling sequence s is the reflection of the juggling diagram of s through a suitable vertical axis. This defines the inverse up to "multiple copies," "cyclic shifts," and "permutations within square brackets." By normalizing suitably as in Subsection 2.6.2, we can define which of all the possible juggling sequences corresponding to the reflected juggling diagram we want to consider as the inverse of s.

- *Scaling.* Given a positive integer m, multiply every integer of a juggling sequence s by m and insert $m - 1$ zeros after every element of the sequence. For example, if we choose $m = 3$ and $s = 31$, we arrive at the new juggling sequence 900300.

- *Concatenation.* If s and t are juggling sequences that correspond to the same juggling state in some state graph, then st is also a juggling sequence that corresponds to the same juggling state.

- *Composition.* We have seen that simple juggling sequences correspond to special permutations of the integers; see Section 2.4. The composition of two permutations that correspond to simple juggling sequences is a new permutation that corresponds to a simple juggling sequence.

- *Vertical Shifts.* If m is the smallest number in a simple juggling sequence, we can subtract any integer less than or equal to m from every element in the juggling sequence to get a new simple juggling sequence; see Section 2.6. Applied to a multiplex juggling sequence, this operation does not, in general, yield another multiplex juggling sequence.

- *Complements.* The complement of a b-ball simple juggling sequence of height h is an $(h - b)$-ball simple juggling sequence of height h; see Subsection 2.8.5 for details about this construction.

- *Unions.* Given two juggling sequences of period p, form the element-wise union to arrive at a new juggling sequence. For example, the union of the juggling sequences 22[72][54][21] and 24[76]42 is the juggling sequence [22][24][7276][544][212]. Note that this also implies that any choice of positive integers enclosed in a pair of square brackets represents a one-element multiplex juggling sequence. Usually, the union of two juggling sequences will be a proper multiplex juggling sequence. However, the union of two simple juggling sequences can also be a simple juggling sequence. For example, the union of 600 and 030 is 630.

4

Multihand Juggling

When we think of a juggler juggling, we usually picture him doing this using both his hands or, sometimes, just one hand. So far, we have been dealing exclusively with patterns that can be juggled using one hand only. In fact, we will consider all the juggling that we encountered so far as 1-hand juggling. In this chapter, we will consider a number of juggling scenarios in which the number of hands involved in performing a juggling pattern becomes important. Note that we do not need to drag multihanded juggling aliens or octopuses into the picture to imagine what juggling with more than two hands would look like. After all, *passing* (that is, juggling balls or other props back and forth among a number of jugglers) is very popular and definitely involves many hands.

4.1 Juggling Matrices

Our first multihand scenario continues where we left off at the end of the last chapter. It generalized 1-hand multiplex juggling, as we considered it there, in a straightforward manner to multihand multiplex juggling. The very basic aspects of this very general kind of multihand juggling were developed by Ed Carstens in the form of his *multihand notation (MHN)* and implemented in his juggling animator *JugglePro*; see [21].

Assuming that you have read and understood the previous chapters, you should be ready to jump straight into the deep end of multihand juggling. Let's consider the fairly typical multihand juggling diagram in Figure 4.1

(the upper diagram). This juggling diagram describes a juggling pattern
of period 4 involving three hands. It should be clear how to interpret this
juggling diagram. In general, as in this example, there will be one row of
beat points per hand, all the catches and throws occur at discrete equally
spaced moments in time, the patterns are periodic, and the number of balls
that get caught at a certain time by a certain hand equals the number of
balls thrown at the same time by the same hand. We'll refer to the hand
that corresponds to the ith row of beat points from the top as hand $i - 1$.
This means that if we are dealing with h hands, then these hands are
numbered from 0 to $h - 1$.

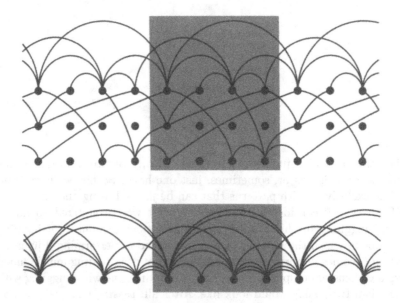

FIGURE 4.1. A juggling diagram corresponding to a 3-hand juggling matrix
of period 4 and the juggling diagram of the corresponding collapsed multiplex
juggling sequence.

In this example, we have three rows of beat points representing the $h = 3$
hands, the period p of the pattern is 4, and, as you can see, there is a lot
of multiplex throwing between hands happening. The appropriate way of
recording such a pattern is no longer a juggling sequence but an h-hand
juggling matrix of period p. This is a $p \times h$ square array. The columns of
this matrix are numbered from 0 to $p-1$, just as the elements of a juggling
sequence of period p are numbered/indexed from 0 to $p-1$. The rows of the
matrix are numbered, just like the hands, from 0 to $h-1$. The (i, j)th entry
of the matrix consists of h sets indexed with the numbers from 0 to $h - 1$,
where set k records the throws that are made at time i from hand j to
hand k. This is the 3-hand juggling matrix of period 4 that corresponds to

our example:

$$\{4,2\}_0\{0\}_1\{2\}_2, \quad \{1\}_0\{0\}_1\{0\}_2, \quad \{2,1\}_0\{0\}_1\{1\}_2, \quad \{3\}_0\{1\}_1\{0\}_2$$
$$\{3,1\}_0\{0\}_1\{0\}_2, \quad \{0\}_0\{0\}_1\{0\}_2, \quad \{0\}_0\{0\}_1\{0\}_2, \quad \{0\}_0\{0\}_1\{0\}_2$$
$$\{4\}_0\{0\}_1\{0\}_2, \quad \{0\}_0\{0\}_1\{0\}_2, \quad \{0\}_0\{0\}_1\{1\}_2, \quad \{0\}_0\{1\}_1\{1\}_2$$

For example, the upper-left entry $\{4,2\}_0\{0\}_1\{2\}_2$—that is, the $(0,0)$th entry—just says that on beat 0, hand 0 performs a 4- and a 2-throw that will end up in hand 0 and a 2-throw that will end up in hand 2. Clearly, if we are dealing with a 1-hand pattern, then we can turn the corresponding juggling matrix into the corresponding multiplex juggling sequence by omitting the subscripts.

Of course, as with multiplex juggling sequences, we also want a compact way of writing juggling matrices. Here is one way of doing this that is in line with what we have done before. If all the sets in an entry contain only 0s, then replace this entry by a 0. Otherwise, omit all the sets that contain 0s, omit all commas outside brackets, replace curly brackets that enclose more than one element by square brackets, and omit curly brackets that contain only one element. Since we have agreed that our examples do not contain throws higher than height 9, also omit all other commas from the matrix. Proceeding in this way, we arrive at the following compact way of writing the juggling matrix above:

$$\begin{array}{cccc} [42]_0 2_2 & 1_0 & [21]_0 1_2 & 3_0 1_1 \\ [31]_0 & 0 & 0 & 0 \\ 4_0 & 0 & 1_2 & 1_1 1_2 \end{array}$$

Again, if we are dealing with a 1-hand pattern, then we can turn the corresponding compact form of the juggling matrix into the corresponding compact form of the multiplex juggling sequence by omitting the subscripts.

Juggling matrices allow us to capture the mathematical essence of just about any juggling pattern that you will ever come across in practice, and they are what juggling animators use as their principal input. Other inputs may include what kinds of props are to be juggled (balls, clubs, chain saws, and so forth), whether or not you want the jugglers to ride unicycles, how many spins (of a club) the different throw heights are supposed to correspond to, and so forth.

Multihand notation really becomes important when we try to record passing patterns involving two or more jugglers. Here are three examples of popular passing patterns involving two jugglers expressed in terms of 4-hand juggling matrices. In these examples, the hands 0 and 1 stand for the right and left hands of the first juggler and hands 2 and 3 for the right and left hands of the second juggler.

$$\begin{array}{cc} 3_3 & 0 \\ 0 & 3_0 \\ 3_1 & 0 \\ 0 & 3_2 \end{array} \qquad \begin{array}{cccc} 3_3 & 0 & 3_1 & 0 \\ 0 & 3_0 & 0 & 3_0 \\ 3_1 & 0 & 3_3 & 0 \\ 0 & 3_2 & 0 & 3_2 \end{array} \qquad \begin{array}{cccc} 7_2 & 0 & 0 & 0 \\ 0 & 0 & 7_3 & 0 \\ 0 & 0 & 0 & 7_1 \\ 0 & 7_0 & 0 & 0 \end{array}$$

The easiest way to juggle these matrices is if you and your partner face each other. The first pattern is "Every Right Hand With Six Balls."[1] On beat 0, both your right hands perform 3-throws to your respective partner's left hands. On beat 1, both your left hands perform 3-throws to your own right hands, and so forth. As far as you are concerned, you do pass–self–pass–self, and so forth.

The second matrix describes the pattern "Every Second Right Hand With Six Balls."[2] It starts out like the first pattern but then continues with self-throws on beats 2 and 3. As far as you are concerned, you do pass–self–self–self–pass–self–self–self, and so forth.

The third pattern[3] is one of the basic patterns with seven balls. Every single throw is a pass, there is one throw on every beat, the hands throw in the sequence Juggler 1 right hand, Juggler 2 left hand, Juggler 1 left hand, Juggler 2 right hand, and so forth. All throws from the right and left hands of Juggler 1 are to the right and left hands of Juggler 2, respectively. In contrast, all throws from the right and left hands of Juggler 2 are to the left and right hands of Juggler 1, respectively.

The 7-throws in the last matrix sound seriously high and, in fact, this last matrix is just the juggling sequence 7 distributed among four hands. However, since the actual action of juggling this sequence is split up among two jugglers such that the juggling is really twice as fast as your own contribution, the actual height at which you and your partner juggle these 7-throws is much less than the height you use for juggling the juggling sequence 7 by yourself in the basic 2-hand cascade pattern.

4.2   Average Theorem and Permutation Test

Most of the results and proofs that we derived for multiplex juggling sequences generalize in a straightforward manner to results about juggling matrices. We only briefly summarize these results in this section.

To start, let's figure out what number of balls is necessary to juggle a pattern described by an h-hand juggling diagram or juggling matrix. As with all the other juggling diagrams, the number of balls equals the number of intersections with the arcs of any vertical line not passing through one of the points of intersection of two arcs. In our first example, this number is 7. The average theorem for juggling matrices runs as follows:

---

[1] file name: two 6 2count.pass in the juggling animator *JoePass!*
[2] file name: two 6 4count.pass in the juggling animator *JoePass!*
[3] file name: two 7 1count.pass in the juggling animator *JoePass!*

The Average Theorem for Juggling Matrices

The number of balls necessary to juggle a pattern corresponding to a juggling matrix equals the sum of the integers in the matrix (excluding the subscripts) divided by its period.

For example, in our complicated first example, the sum of all integers in the juggling matrix is 28 and, since the period of the matrix is 4, we confirm that we are really juggling 28/4=7 balls.

The proofs of the above and the following result are straightforward extensions of the proofs of the respective results for multiplex juggling sequences and will be omitted here.

Let $s = \{S_{i,j} \mid i = 0, 1, \ldots, p - 1, j = 0, 1, \ldots, h - 1\}$ be a $p \times h$ array such that all entries $S_{i,j}$ are built like those of an h-hand juggling matrix of period p (compact or full). We transform s into a new matrix s' by doing the following:

- First, delete all 0s from its entries, and delete all commas if you are dealing with the full matrix.

- Then, provide the remaining integers in brackets with the subscript of the enclosing brackets, and remove the brackets themselves and their subscripts.

- Given an integer a (not a subscript) in column i, replace it by the integer $(a + i) \bmod p$.

For example, if s is the (4×3) matrix

$$
\begin{array}{lll}
[42]_0 2_2 & 1_0 & [21]_0 1_2 & 3_0 1_1 \\
[31]_0 & 0 & 0 & 0 \\
4_0 & 0 & 1_2 & 1_1 1_2
\end{array}
$$

that we considered at the beginning of this chapter, then s' is

$$
\begin{array}{lll}
0_0 2_0 2_2 & 2_0 & 0_0 3_0 3_2 & 2_0 0_1 \\
3_0 1_0 & & & \\
0_0 & & 3_2 & 0_1 0_2
\end{array}
$$

Now, you can double-check, using the following result, that our sample (4×3) matrix is really a juggling matrix:

The Permutation Test for Juggling Matrices

Let s be a $(p \times h)$ matrix all of whose entries are built like those of an h-hand juggling matrix of period p (compact or full), and let s' be the $(p \times h)$ matrix constructed from s as detailed above. Then, s is an h-hand juggling matrix of period p if and only if for all $i = 0, 1, \ldots, p-1$, $j = 0, 1, \ldots, h-1$ the number of integers (not subscripts) in the (i, j)th entry of s' equals the total number of integer/subscript pairs i_j in the matrix s'.

4.3   Multihand State Graphs

At this stage, it should also be clear how to define and construct the *b-ball h-hand multihand state graphs of height k*, with $b, h, k \geq 1$, that can be used to find all *b-ball h-hand juggling matrices of height k*.

This graph has

$$\binom{b + hk - 1}{b}$$

vertices. This formula can be derived using the same ideas that we used to derive the corresponding formula for multiplex state graphs. Figure 4.2 shows the 2-ball 2-hand multihand state graph of height 2. We did not label the edges with multihand throws because things are very crowded even without these labels. Some more special features of multihand state graphs show up in our example. Note, for example, that there are two different ways to move from the state at the top of the diagram to some of the other states. Also, it is no longer clear which state should be called the ground state, and we would have to spend some more time setting the stage if we were interested in defining necklaces for this graph.

As in the case of multiplex state graphs, it is not hard to show that the states a juggling matrix visits determine the throws in the matrix.

States Determine Throws

If s and s' are two b-ball h-hand juggling matrices in compact form that visit the same states in the b-ball h-hand multihand state graph of some height the same number of times each, then s contains the same throws the same number of times as s'.

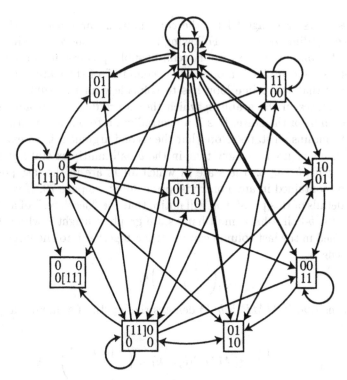

FIGURE 4.2. The 2-ball 2-hand multihand state graph of height 2 with edge labels omitted.

For example, the two juggling matrices

$$
\begin{array}{cccc}
1_0 & 2_1 & 2_0 & 0 \\
2_0 & 0 & 0 & 1_1
\end{array}
\qquad
\begin{array}{cccc}
2_0 & 0 & 2_1 & 1_0 \\
1_1 & 2_0 & 0 & 0
\end{array}
$$

visit the same states in Figure 4.2 starting with the state at the top. As you can see, both matrices really contain three 0s, two 1s, and three 2s each.

The circular path in Figure 4.2 corresponds to a *maximal prime loop* in this graph. The *maximal prime juggling matrix* that corresponds to this loop and the starting state at the top is

$$
\begin{array}{ccccccccc}
1_0 & 2_1 & 2_1 & 0 & 0 & [22]_1 & 0 & 0 & 0 \\
2_0 & 0 & 0 & 2_0 & 1_0 & 0 & 0 & 2_0 2_1 & 0
\end{array}
$$

Note that the highlighted maximal prime loop contains all vertices of the graph except one. In general, it is possible to prove that maximal prime loops of the b-ball h-hand multihand state graph of height k, with $b, h, k \geq 2$, cannot include all the vertices of the graph. We only sketch a proof for this result. Assume there was a prime loop that contains all states. Then, we can use the same idea that is needed to prove the throws-from-states result above to conclude that all throws in a corresponding juggling matrix are either 0s or ks; see page 54 for the corresponding argument in the case of simple juggling sequences. Following this, we can show that the prime loop has to contain at least one nonempty subloop consisting of states all of whose 1s are contained in one of the "columns" of the state. Of course, this means that the prime loop has to coincide with this subloop. Since there are states in the graph in which the 1s are distributed among different columns of the state, this is a contradiction to our assumption that the loop contains all states. Note that the second argument is an extension of the argument used to show that in the b-ball multiplex state graph of height k, with $b, k \geq 2$, the state in which all 1s are in the last position cannot be contained in any nontrivial prime loop; see page 79. If you work out the details of this proof, you will notice that the "necklaces" of all those states of the b-ball h-hand multihand state graph of height k whose 1s are all contained in the last column can be strung together into one large prime loop of length

$$\binom{b+h-1}{b} k.$$

This means that if $MMP(b, h, k)$ denotes the length of a maximum prime loop, then

$$\binom{b+h-1}{b} k \leq MMP(b, h, k) \leq \binom{b+hk-1}{b} - 1.$$

Note that if you set the number of hands to 1 while leaving $b, k \geq 2$, these inequalities stay valid, as in this case $MMP(b, h, k)$ turns into $MM(b, k)$, the length of a maximal prime loop in the b-ball multiplex state graph of height k, and the inequalities turn into inequalities that we derived in Section 3.4 for $MM(b, k)$.

Nobody has looked into enumerating juggling matrices, and we will not pursue this line of research any further either. This is mainly because, as we have seen in the case of multiplex juggling sequences, with the available methods, we are not really able to calculate the numbers that we are interested in. However, it is certainly a fairly straightforward exercise to define multihand juggling cards and, using these, to calculate upper bounds for the number of b-ball h-hand juggling matrices of period p. It is also possible to use the idea of expanding multiplex juggling sequences into multihand juggling matrices, introduced in the following section, to calculate such upper bounds based on the upper bounds for the number of b-ball multiplex juggling sequences of period p that we calculated before.

4.4   Operations Involving Juggling Matrices

The following operations to transform multiplex juggling sequences into new multiplex juggling sequences described in Section 3.5 also all have counterparts for juggling matrices: forming multiple copies, cyclic shifts, permutations within square brackets, site swaps, inverses, scalings, concatenations, and unions. There are some additional operations that can be used to transform juggling matrices into new juggling matrices. For example, just renumbering the hands results in a different juggling matrix. In terms of juggling matrices, renumbering of hands translates into a permutation of the rows of the matrix and a corresponding adjustment of indices.

Given a juggling diagram of a multihand pattern, we can *contract* all beat points that correspond to the same beat into one beat point. The resulting diagram is a (1-hand) multiplex juggling diagram. We call it the *contracted juggling diagram*. The second diagram in Figure 4.1 shows the contracted juggling diagram of our first complicated multihand juggling diagram. Of course, we can also perform this operation at the level of juggling matrices. Given such a matrix, we first omit all indices. Then, we combine all the sets in one column into a single set, and, if this set contains integers other than 0s, we omit all 0s from the set, or, if this set contains only 0s, we omit all except one of these 0s. By proceeding in this way, we arrive at the multiplex juggling sequence that corresponds to the contracted juggling diagram. We call this juggling sequence the *contracted juggling sequence* of the juggling matrix with which we started. In our first example, we get

$$\{4, 2, 2, 3, 1, 4\}, \{1\}, \{2, 1, 1, 1\}, \{3, 1, 1, 1\},$$

or, for short,

$$[422314]1[2111][3111].$$

It turns out that there is an easy systematic procedure to construct, starting with an arbitrary multiplex juggling diagram, all h-hand juggling diagrams that contract onto this multiplex juggling diagram. We call the basic operation on which this procedure is based a *side step*. Given a multihand juggling diagram of period p, a side step consists in the following (see Figure 4.3 for an example of a side step). Start by choosing a beat point that has at least one arc coming in from the left. For example, in the diagram, choose one of the open points. Then, choose one of the incoming arcs and one of the outgoing arcs in this beat point (let's choose the ones drawn as double lines) and grab the two end points of these two arcs that coincide with the beat point plus all corresponding (via the period of the pattern) end points of arcs. Finally, move all the end points that you have selected up or down until they coincide with other beat points.

Now, given an h-hand juggling diagram, you can use a finite number of side steps to turn it into an h-hand juggling diagram in which only one of the hands is doing something. By omitting all the beat points corresponding to hands that do nothing, we arrive at the contracted multiplex juggling diagram. On the other hand, starting with a multiplex juggling sequence, we can systematically construct all juggling matrices that contract to this sequence using side steps.

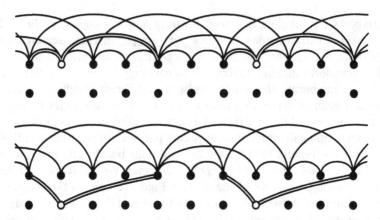

FIGURE 4.3. Performing a side step on a 2-hand juggling diagram gives another 2-hand juggling diagram.

Side steps are very well-behaved operations on juggling matrices in that they preserve the number of balls and hands necessary to juggle a pattern and the period of the pattern.

4.5   Special Classes of Juggling Matrices

The results and techniques described so far in this and the previous chapters have counterparts and can be brought to bear on many well-defined and interesting classes of multihand juggling matrices.

For the sake of completeness, we first note that 1-hand juggling matrices are equivalent to multiplex juggling sequences and "simple" 1-hand juggling matrices are equivalent to simple juggling sequences.

A juggling matrix is *simple* if on every beat every hand handles at most one ball. To get from the concept of simple juggling sequences to that of juggling matrices, we first introduced multiplex throws and then multiple hands. It would have been equally natural to first introduce multiple hands, yielding the simple multihand matrices, and then multiplex throws. In fact, the majority of multihand juggling matrices performed by jugglers are simple.

A juggling matrix is *distributed* if on every beat at most one of the hands is doing something. Basically, a distributed juggling matrix is a juggling sequence "distributed among different hands." In fact, most juggling animators and jugglers will automatically interpret a simple or multiplex juggling sequence as a 2-hand juggling matrix. For example, hardly any juggler will try to perform the simple juggling sequence 441 using one hand only. Instead, what jugglers will really do is interpret it as the 2-hand juggling matrix

$$\begin{matrix} 4_0 & 0 & 1_1 & 0 & 1_1 & 0 \\ 0 & 1_0 & 0 & 4_1 & 0 & 1_0 \end{matrix}$$

in which the hands take turns performing the throws in the sequence. Note that since 441 has an odd period, we have to double up the sequence before we can turn it into a juggling matrix. Furthermore, odd-height throws are made to cross from one hand to the other while even-height throws are performed as self-throws.

A *cyclic juggling matrix* is a special distributed juggling matrix in which the only hand that can possibly do something on beat i is hand $i \bmod h$. Furthermore, the period p of a cyclic juggling matrix is a multiple of the number of hands h. For example, the default 2-hand interpretation of a juggling sequence is a cyclic 2-hand juggling matrix. It should be clear that any juggling sequence can be distributed into many different h-hand juggling matrices for any $h \geq 2$ (cyclic and otherwise). For example,

$$\begin{matrix} 4_0 & 1_1 & 0 & 0 & 1_0 & 1_1 & 0 & 0 & 1_0 & 4_0 & 0 & 0 \\ 0 & 0 & 1_1 & 4_1 & 0 & 0 & 4_1 & 1_0 & 0 & 0 & 1_1 & 1_0 \end{matrix}$$

describes a second very different way to juggle 441 using two hands. We will continue this line of thought in Section 4.9.

A straightforward way of building a (simple) h-hand juggling matrix of period p is to combine h (simple) juggling sequences of period p by having the different hands juggle the different juggling sequences independently from each other. For example, the two juggling sequences 313131 and 441441 of period 6 combine into the 2-hand juggling matrix

$$\begin{matrix} 3_0 & 1_0 & 3_0 & 1_0 & 3_0 & 1_0 \\ 4_1 & 1_1 & 1_1 & 4_1 & 1_1 & 1_1 \end{matrix}$$

The most basic 2-hand juggling matrices built like this are the 1-column matrices

$$\begin{matrix} n_0 \\ n_1 \end{matrix}.$$

For example, in the case of $n = 2$, we juggle four balls: two balls in the left hand and two balls in the right hand. Of course, we can also make balls cross as in

$$\begin{matrix} n_1 \\ n_0 \end{matrix},$$

or we can mix self- and crossing throws as in

$$\begin{matrix} n_0 & n_1 \\ n_1 & n_0 \end{matrix}.$$

These matrices are also the juggling matrices of what jugglers usually refer to as *simultaneous* (2-hand) patterns. These are patterns in which on at least one of the beats two hands catch and throw at the same time. In fact, in the basic simultaneous patterns, all hands perform throws on every beat.

The exact way of noting juggling matrices varies widely and, for the most popular classes of juggling patterns, more streamlined ways of noting the patterns belonging to these classes are used and are accepted as input by many of the juggling animators. As we have already mentioned, juggling sequences are usually interpreted as cyclic 2-hand juggling matrices. Furthermore, most jugglers will perform an n-throw in a simultaneous (2-hand) pattern at the same height as a $(2n)$-throw in a cyclic (2-hand) pattern. Because of this, the way the basic simultaneous pattern above is usually noted is $(2n, 2n)$. More complicated patterns such as

$$3_1 \quad 3_1 \quad 1_0$$
$$2_1 \quad 3_0 \quad 3_0$$

are noted in the form

$$(6x, 4)(6x, 6x)(2, 6x).$$

Here, we have one pair of brackets per column, every pair of brackets has two entries, and the first and second entries describe what hands 0 and 1 do. If an entry is just a number, it represents a self-throw, and if it is followed by an x, it represents a crossing throw.

4.6   Uniform Juggling and Shannon's Theorems

In [118], Claude Shannon proved the first juggling-related theorems in the history of juggling and mathematics. In the following, we state and prove these results.

Unlike earlier in this chapter, throws are not necessarily required to occur at distinct equally spaced moments in time (at least not to start with), and juggling is no longer only performed in hot-potato style; that is, the time a ball is held by a hand between catching and releasing it again is taken into account.

Let's get things moving and assume that you and your friends are using h hands to juggle b balls in the manner detailed above (both h and b being nonzero), and you have been doing this forever and will be doing this forever.

You are juggling *uniformly* if and only if at all times the following *synchronicity properties* hold:

1. The *dwell time* of a ball (or a hand) is a constant d; that is, the time that any ball is held by any hand between catching and throwing is d.

2. The *flight time* of a ball is a constant f; that is, the time that any ball spends in the air between being thrown and being caught again is f.

3. The *vacant time* of a hand is a constant v; that is, the time that any hand spends vacant between throwing and catching again is v.

In uniform juggling, there is a nice *duality* between balls and hands. To understand this, let's have a closer look at a group of jugglers who are juggling uniformly. Let's imagine that the jugglers only consist of hands. This means that what you see when you witness uniform juggling in action is only hands and balls flying around madly. Now, turn the world upside-down and at the same time turn hands into balls and balls into hands (the juggling is still continuing). After this transformation, you can check that what you are witnessing now (the right side up) is uniform juggling with b hands and h balls, flight time v, vacant time f, and dwell time d. What all this says is that while we are juggling uniformly we are juggling both balls and hands in the same sense. This implies that, given any true statement about uniform juggling, we can immediately translate this true statement into another true statement about uniform juggling.

Principle of Duality

Any true statement about uniform juggling turns into another true statement about uniform juggling by replacing any word, phrase, or parameter associated with "balls" by the respective word, phrase, or parameter associated with "hands" and vice versa.

For example, by applying this principle to Property 2 above, we arrive at Property 3. Applying the principle to Property 1 leaves this statement unchanged. This means that Property 1 is *self-dual*. Of course, by applying the duality statement twice in a row, we end up with the statement we started with. We only remark that principles of duality such as the above play important roles in a number of mathematical disciplines. For example, in projective geometry, points and hyperplanes of projective spaces are dual objects very much like hands and balls are dual objects in uniform juggling.

It is probably fair to say that most jugglers will spend most of their juggling lives picking up dropped props and trying to juggle uniformly. For example, actually performing any of the one-element simple juggling sequences with one hand will be done uniformly. Performing the same sequences with two hands (as cascades and fountains), with the two hands taking turns, will also be done uniformly. In addition, for two hands and an even number of balls, there are many different ways to juggle uniformly. For example, consider the case of the 2-ball fountain. Here, one ball is manipulated by the left hand and the other by the right hand. Now, instead of taking turns throwing balls, the two hands can throw simultaneously.

Every time we throw, we have a choice between two simultaneous throws that cross to the respective other hand and two throws every one of which ends in the same hand that is throwing it. This means that even in this very simple setup there are as many "different" ways to juggle uniformly as there are 0-1 sequences; that is, infinitely many. However, if we do not distinguish between simultaneous crossings or self-throws, then there are only two different ways to uniformly juggle two balls with two hands: alternating or simultaneous. The same is true for any even number of balls and two hands. More generally, we will find that for any possible combination of balls and hands there are only finitely many essentially different ways to juggle uniformly.

There is a general relationship discovered by Claude Shannon that ties together the five parameters b, h, d, f, and v. Note that this relationship is self-dual.

Shannon's First Theorem

In uniform juggling with b balls, h hands, dwell time d, flight time f, and vacant time v, we have

$$\frac{f+d}{v+d} = \frac{b}{h}.$$

Of course, if we want to use this equation in real-life juggling, we have to be able to conclude that if a juggle has been proceeding for a sufficiently long period of time such that during that period the synchronicity conditions hold, then the equation above holds. By inspecting the following proof, you will find that long enough is $h(f+d)$; that is, the time it would take a ball to visit all hands in sequence. Also, once a juggle has been going on for such a period of time, it is part of one of our ideal uniform juggles that extends forever into the past and future.

Let's prove Shannon's theorem. We first concentrate on a ball that gets thrown at some point in time sufficiently far back in the past that it has been caught at least h times since then. Because there are only h different hands and the ball has visited $h + 1$ by now, it must have visited one of the hands at least twice. Let's focus on a hand that was visited twice. Then, between the first and the second times that the ball was caught by this hand, the ball was caught another m times. Using Properties 1 and 2 above, we conclude that between the first catch and the second catch of the ball by the distinguished hand, $(m+1)(f+d)$ of time has passed. Also, between the first catch and the second catch, the distinguished hand has made n further catches. Using Properties 1 and 3, we conclude that this took $(n+1)(v+d)$ of time and that therefore

$$(m+1)(f+d) = (n+1)(v+d).$$

Let $p = (m+1)/g$ and $q = (n+1)/g$, where g is the greatest common divisor of the two integers $m+1$ and $n+1$. This means that

$$p(f+d) = q(v+d) \tag{$*$}$$

and that p and q are relatively prime.

To prove the first theorem, consider a time interval of length $p(f+d)$ such that no ball is being caught at the beginning of this interval. Balls get caught within this interval at times t_1, t_2, t_3, and so forth. At time t_i, we find that exactly s_i hands catch s_i balls. Every one of our balls gets caught exactly p times in the distinguished interval, and every hand makes exactly q catches. We conclude that

$$\sum s_i = pb = qh. \tag{$**$}$$

We arrive at the desired result by combining this equation with equation $(*)$.

We proceed to figure out how exactly things get juggled when we are juggling uniformly and that, in a certain sense, there are only a finite number of different ways to uniformly juggle b balls with h hands. We will also derive an explicit formula for this number.

Continuing our argument from above, instead of focusing on one ball, let's see what happens to all the balls that get thrown at a certain time u_0; let's say there are v_0 of them. Because of the synchronicity properties, these balls will always be caught and thrown at the same time as long as we are juggling uniformly. Furthermore, no other balls will get caught when they are caught, and no other balls will ever be thrown when they are thrown. From equation $(*)$, it now follows that exactly the same set of hands we started with will catch them the pth time they are caught, and this is the first time after u_0 that this happens.

Now, consider the set of hands that threw at time u_0. The same set of hands will again throw exactly v_0 balls at time $u_0 + v + d$. This second set of balls will visit exactly the same sets of hands in the same order as our first set. Continuing this argument, we conclude that there are q sets of v_0 balls following each other cyclically, visiting the same p sets of v_0 hands each. This means that there are altogether qv_0 balls being manipulated by pv_0 hands and that this system of balls and hands is completely independent; none of the balls in this system interacts with any of the hands outside the system and none of the hands with any of the balls outside the system. This means that if we forget about the balls and hands outside the system, we see uniform juggling with qv_0 balls and qv_0 hands.

If the qv_0 balls did not exhaust all of the balls, we carry out the same process with another set of v_1 balls that get thrown at a time different from the times that any of the qv_0 balls gets thrown. This yields a set of qv_1 balls that is disjoint from the first one. We continue like this until all balls have been covered in k steps. Hence,

$$b = q \sum_{i=0}^{k-1} v_i \text{ and } h = p \sum_{i=0}^{k-1} v_i. \qquad (***)$$

Since p and q are relatively prime, we conclude that $\sum_{i=0}^{k-1} v_i$ is the greatest common divisor of b and h and that what we are looking at is the "disjoint union" of k independent hand–ball systems (groups of jugglers) every single one of which performs uniform juggling.

Shannon's second theorem is an immediate consequence of all this.

Shannon's Second Theorem

Uniform juggling with b balls and h hands, with b relatively prime to h, can be done in a uniquely determined way (up to labeling). The balls can be numbered from 0 to $b-1$ and the hands from 0 to $h-1$ in such a way that each ball progresses through the hands in cyclical order and each hand catches the balls in cyclical order.

To draw a picture of this essentially unique way to uniformly juggle b balls with h hands, with b relatively prime to h, start with the juggling diagram of the constant b-ball juggling sequence and then label the beat points cyclically from 0 to $h-1$ and the orbits (=balls) cyclically from 0 to $b-1$. Finally, split up the beat points into a catch and a throw to account for nonzero dwell time. Note that the dwell time does not really matter in terms of the overall structure of this pattern, and we may as well have chosen to leave this dwell time equal to zero.

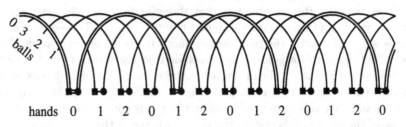

FIGURE 4.4. Juggling four balls using three hands uniformly.

Figure 4.4 illustrates this in the case of four balls and three hands. Follow the orbit of ball 0 (drawn with double lines) to convince yourself that these balls really progress through the hands in cyclical order (... , 0,1,2, ...).

Similarly, convince yourself that hand 0 catches the balls in cyclical order. Note that the dual of this unique way of uniformly juggling four balls with three hands is the unique way of uniformly juggling three balls with four hands.

Now, it is time to translate equation (∗ ∗ ∗) and the remarks leading up to it into Shannon's third theorem.

Shannon's Third Theorem

Uniform juggling with b balls and h hands, with g being the greatest common divisor of b and h, is possible in as many essentially different ways (we specify this below) as there are representations of g as a sum of positive integers. We call these essentially different ways of juggling uniformly the *uniform juggling patterns with b balls and h hands*.

For example, consider again the case of two hands and an even number of balls. Then, the greatest common divisor of b and h is 2, and 1+1 and 2 are the only ways to write this greatest common divisor as the sum of positive integers. Here, 1+1 corresponds to the alternating fountain pattern (the throws with the left hand are performed at different times from those with the right hand and are performed one at a time). On the other hand, 2 stands for the simultaneous fountain pattern in which two balls are thrown every time a throw is made.

In general, assume there are h hands and b balls, let g be the greatest common divisor of b and h, $p = b/g$ and $q = h/q$ just as above, and let

$$g = \sum_{i=0}^{k-1} v_i$$

be a representation of g as a sum of positive integers. Then, a uniform juggling pattern that corresponds to this representation consists of k independent uniform juggling patterns J_i, $i = 0, 1, \ldots, k-1$ in which $v_i q$ balls are juggled by $v_i p$ hands, v_i balls at a time, and that share the same dwell, flight, and vacant times. Furthermore, given juggling patterns J_i and J_j, with $i \neq j$, the times at which balls get tossed in J_i never coincide with times at which balls get tossed in J_j.

This also specifies what we mean by "essentially different"—namely different up to relative displacements of the different juggling patterns J_i in time, choice of dwell, flight and vacant times, and shuffling of the balls at every throw within one of the J_is.

Table 4.1 shows the number of sum representations for some small values of g.

g	1	2	3	4	5	6	7	8	9	10
sum representations of g	1	2	3	5	7	11	15	22	30	42

TABLE 4.1. The number of representations of the positive integer g as a sum of positive integers for small values of g.

To conclude this section, let us return to the first theorem. Pick one of the different uniform juggling patterns with b balls and h hands, $b > h$. Suppose you want to juggle this pattern using a fixed flight time f. Then, the first theorem allows us to calculate the possible range of the *frequency* $v + d$ of one of the hands throwing. Despite f being fixed, we can change this period by varying vacant time v and dwell time d. We have

$$v + d = \frac{(f+d)h}{b} = \frac{(f-v)h}{b-h}.$$

This means that the frequency will be minimal for $d = 0$ and maximal for $v = 0$. The ratio of these two extremal values only depends on b and h.

The Frequency of a Uniform Juggling Pattern

The frequency of a uniform juggle with b balls and h hands, $b > h$, and fixed flight time can be varied by changing the dwell and vacant times. The ratio of the maximum and minimum possible frequencies is

$$\frac{b}{b-h}.$$

For example, with the 3-ball cascade, this ratio is 3 to 1, with the 5-ball cascade 5 to 3, and with the general $(2n+1)$-ball cascade $2n+1$ to $2n-1$. This means that with larger numbers of balls and fixed flight time, less and less range is available. In real juggling, the possible ranges will be considerably smaller than suggested by these ratios since it is not possible to have zero vacant time or zero dwell time.

A uniform juggling pattern can be described by a juggling matrix if all throws occur at discrete equally spaced moments in time. From our considerations above, it follows that a juggling matrix describes a uniform juggling pattern if and only if it has the following properties:

- The matrix is simple.

- Every entry of the matrix is either equal to 0 or equal to a certain fixed positive integer u.

- There is a constant c such that the number of 0s between any pair of consecutive us in any of the rows equals c. Furthermore, the number of 0s before the first u in a row plus the number of 0s behind the last u in a row equals c.

4.7   Shannon's Theorems for Juggling Sequences

In [155], Jinjang Song and Yeung Yam proved a counterpart of Shannon's first theorem and some corollaries in the framework of juggling sequences. In the following, we summarize, correct, and complement their results.

We start with a (simple) juggling sequence $a_0 a_1 \cdots a_{p-1}$ of period p. Then, we introduce multihandedness as well as dwell, flight, and vacant times into the juggling of this sequence such that the following hold:

- Catches are made on beats.

- We juggle using h hands (denoted 0 to $h-1$) such that h divides p and the hands take turns throwing the balls in cyclical order. So, what we are doing is interpreting the juggling sequence as an h-hand cyclic juggling matrix.

- For an a_i-throw, the dwell and flight times are $a_i \omega_i$ and $a_i(1 - \omega_i)$, respectively, where ω_i is chosen such that $0 \leq \omega_i \leq 1$. If we juggle a juggling sequence without 0s in this way, one hand makes a catch every h beats. Therefore, to avoid more than one ball ending up in one of the hands, we also require that all dwell times be less than h; that is, we require that

$$a_i \omega_i < h, i = 0, 1, 2, \ldots, p - 1.$$

  Note that this is automatically satisfied if $a_i < h$. (We only remark that for juggling sequences that contain zeros it would sometimes be possible to allow for longer dwell times if our only aim was to avoid two balls in one hand. For example, in the 3-ball juggling sequence 900 juggled with two hands, we could choose any number between 0 and 6 as the dwell time for the 9-throws. A detailed analysis of what is possible in this respect is messy and does not seem to be worth the additional effort.)

Figure 4.5 shows the picture that you should keep in mind in the case of the juggling sequence 31 juggled with two hands. The upper diagram is basically the juggling diagram of 31. The only difference is that the left- and right-hand beats are placed at slightly different levels. The lower diagram is the corresponding juggling diagram that accounts for nonzero dwell time. The period of the juggling sequence 31 is 2. One such period is highlighted in gray. Every period is supposed to start on a beat.

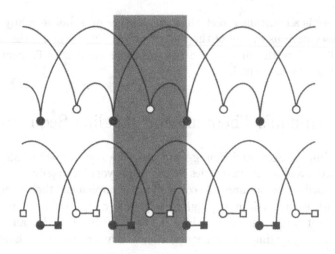

FIGURE 4.5. Juggling the juggling sequence 441 with two hands with zero and nonzero dwell times.

To calculate the *total flight time* f_T of all the balls during one period, note that there is a complete flight path of a 1-throw (a_1-throw) contained in the highlighted period. This arc accounts for one ball being in the air for a time of $a_1(1 - \omega_1)$. Furthermore, there are two parts of an arc representing 3-throws (a_0-throws) contained in the period. These two parts can be combined into one full arc. This arc accounts for a flight time of $a_0(1 - \omega_0)$. In the general case, it can be shown that the parts of arcs representing a_i-throws within a period combine into exactly one such full arc (here, different a_is with the same value are considered to be different). Hence,

$$f_T = \sum_{i=0}^{p-1} a_i(1 - \omega_i).$$

Similar considerations yield that the *total dwell time* d_T of all the hands during the highlighted period is

$$d_T = \sum_{i=0}^{p-1} a_i \omega_i.$$

Finally, the *total vacant time* v_T for all hands is clearly

$$v_T = hp - d_T.$$

These expressions for the total dwell, flight, and vacant times combine into

$$\frac{f_T + d_T}{v_T + d_T} = \frac{\sum_{i=0}^{p-1} a_i(1 - \omega_i) + \sum_{i=0}^{p-1} a_i \omega_i}{hp - d_T + d_T} = \frac{\sum_{i=0}^{p-1} a_i}{hp} = \frac{b}{h}.$$

Note that in the last step of manipulating this equation, we used the Average Theorem for juggling sequences; see page 15.

Shannon's First Theorem for Juggling Sequences

We are juggling a juggling sequence with fixed dwell, flight, and vacant times, as specified above, such that summed over a period of this juggling sequence the total dwell, flight, and vacant times of all balls and hands are d_T, f_T, and v_T. Then,

$$\frac{f_T + d_T}{v_T + d_T} = \frac{b}{h}.$$

Let's say we are watching a group of jugglers perform a juggling sequence with period p. We want to be able to conclude that the equation in the theorem holds during some fixed period that starts at time t_0 and ends at time $t_0 + p$. For how long do we have to check that the sequence is juggled according to the rules set out above to be able to make this conclusion? If a_{max} is the maximum throw height in the sequence and ω_{max} the maximum value among the ω_is, then any throw that has anything to do with this period is caught and launched within the time interval

$$]t_0 - a_{max} - a_{max}\omega_{max}, t_0 + p + a_{max}\omega_{max}[.$$

Now, it is clear from the proof that if we check that the juggling proceeds as prescribed during this interval, then the equation in the theorem holds within the period in which we are interested.

We have seen that the frequency $(=v + d)$ of a uniform b-ball juggling pattern with h hands and fixed flight time can be varied by changing the dwell time d and vacant time v and that the ratio of the maximum and minimum possible frequencies is $b/(b - h)$; see page 102. This result is a corollary of Shannon's first theorem.

From the way we set up things, it is clear that when we are performing a juggling sequence $a_0 a_1 \cdots a_{p-1}$, we are always juggling with *total frequency* $v_T + d_T = hp$. However, the total flight time can be changed by varying the dwell times. To be able to derive the counterpart of the corollary to Shannon's first theorem, we could slow down or speed up things as we vary the dwell times such that the total flight time stays fixed. Of course, the ratio of the maximum total frequency and the minimum total frequency in the fixed total flight time setting equals the ratio of the maximum total flight time and the minimum total flight time in the fixed total frequency setting. Let's calculate the latter one.

Remember that the total flight time is

$$f_T = \sum_{i=0}^{p-1} a_i - \sum_{i=0}^{p-1} a_i \omega_i.$$

This implies that we get a maximum total flight time by choosing all dwell times to be zero, that is,

$$f_{Tmax} = \sum_{i=0}^{p-1} a_i.$$

Since $a_i \omega_i < h$ and $0 \leq \omega < 1$, we get a minimal total flight time by setting $a_i \omega_i = h$ if $a_i > h$ and equal to a_i otherwise. Hence, the minimum total flight time

$$f_{Tmin} = \sum_{\{i|a_i>h\}} (a_i - h).$$

This means that the ratio we are after is

$$\frac{f_{Tmax}}{f_{Tmin}} = \frac{\sum_{i=0}^{p-1} a_i}{\sum_{\{i|a_i>h\}} (a_i - h)}.$$

Note that this ratio can by infinity if $f_{Tmin} = 0$. In the case of the one-element juggling sequence b such that $b > h$, we are dealing with a uniform juggle. In this special case, the period p equals 1 and the ratio above turns into the ratio

$$\frac{b}{b - h}$$

predicted by the corollary to Shannon's first theorem. We now formulate the analog of this corollary within the framework under consideration. We only note that in [155] a similar result is proved which, however, does not quite count as an analog.

The Frequency of Juggling a Juggling Sequence

We are juggling a juggling sequence $a_0 a_1 \cdots a_{p-1}$ with h hands. Then, in the fixed total flight time setting, the ratio of the maximum total frequency and the minimum total frequency is

$$\frac{\sum_{i=0}^{p-1} a_i}{\sum_{\{i|a_i>h\}} (a_i - h)}.$$

From the equation $v_T + d_T = hp$, we see that the total frequency is really a measure of how fast we juggle one whole period of the juggling sequence.

Therefore, the ratio above is also the ratio of the fastest and slowest possible ways of juggling the juggling sequence (in the fixed total flight time setting). For example, the ratio for the 3-ball juggling sequence 51 with two hands is $(5+1)/(5-2) = 2$. This means that if we juggle 51 with zero dwell times, we juggle twice as fast as if we juggle it with maximum dwell times.

It should be possible to prove results similar to those developed in this section in the more general setting in which both the balls and the hands juggle nontrivial juggling sequences; see Section 4.9 for more details about this more general setting.

Shannon's second and third theorems deal with the natural decomposition of uniform juggling patterns into independent uniform juggling patterns. We will discuss the corresponding problem for juggling sequences in Section 4.11.

4.8   Cascades and Fountains

After juggling with dwell time in the previous two sections, we'll again be juggling without dwell time for the rest of this chapter.

Juggling the basic b-ball juggling sequence with two hands leads to cascades and fountains for odd and even b, respectively. In the following, we will consider the multihand counterparts of cascades and fountains. This material first appeared in Wolfgang Schebeczek's article [111]; see also [37].

The following definitions are natural generalizations of the 2-hand case. The *basic b-ball juggling pattern with h hands* is what you get when you juggle the basic b-ball juggling sequence with h hands (denoted 0 to $h-1$) such that the hands take turns throwing the balls in cyclical order; see Figure 4.6 for the picture to keep in mind illustrated in the case of three balls and six hands. The juggling matrix of a basic b-ball juggling pattern with h hands is cyclic.

We call such a pattern a *cascade* if all hands are touching all balls. If, in addition, each hand throws to the hand that is throwing next, then we call it a *next-hand cascade*. Similarly, a cascade is a *previous-hand cascade* if each hand throws to the hand that precedes it in the order of throwing. All basic b-ball juggling patterns with one hand are cascades. Furthermore, in the 1- and 2-hand cases, all cascades are both next-hand and previous-hand cascades.

We call one of the patterns under consideration a *fountain* if it is not a cascade, or equivalently, if the pattern splits into at least two independent patterns. We call a fountain a *same-hand fountain* if all throws by any hand are to itself. In the 2-hand case, all fountains are same-hand fountains.

We first observe that the basic b-ball juggling pattern with h hands is a realization of the uniform b-ball juggling pattern with h hands that corresponds to the sum representation all of whose summands are 1s; see

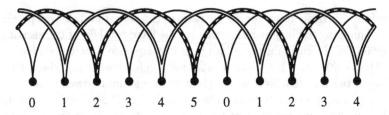

0   1   2   3   4   5   0   1   2   3   4

FIGURE 4.6. The basic 3-ball pattern with six hands labeled 0, 1, 2, 3, 4, and 5 splits into three independent patterns in which two hands juggle one ball each. It is a fountain but not a same-hand fountain.

Shannon's third theorem in Section 4.6 ("realization" in the sense that dwell time is zero, flight time is b, vacant time is h, and the hand throws are arranged to occur in cyclical order and at equally spaced moments in time that are one unit of time apart). We conclude immediately that the dual of the basic b-ball juggling pattern with h hands is the basic h-ball juggling pattern with b hands; see Section 4.6 for a definition of the term dual. This also means that in the following every result about balls can be translated into a result about hands as prescribed by the duality principle.

By applying Shannon's theorems (see Section 4.6), we immediately derive the following characterization of multihand cascades and fountains:

Multihand Cascades and Fountains I

The basic b-ball juggling pattern with h hands is a cascade if and only if b and h are relatively prime and a fountain otherwise. If g is the greatest common divisor of b and h, then the pattern splits into g independent patterns whose common juggling sequence is

$$b\underbrace{00\cdots0}_{g-1 \text{ times}} .$$

In each of these subpatterns, h/g hands are juggling b/g balls.

If b and h are not relatively prime, then the g subpatterns of the basic b-ball juggling pattern with h hands are best described as "slow" basic (b/g)-ball juggling patterns with h/g hands. These are "slow" cascades since b/g and h/g are relatively prime.

A basic pattern with just one more ball than hands is called a *domino pattern* because a throw from some hand is always made to the next hand in the order of throwing and, in turn, causes this next hand to make the next throw, very much in analogy with one falling domino causing its neighbor to fall. Domino patterns are next-hand cascades. In general, it is easy to verify the following corollary of the result above:

Multihand Cascades and Fountains II

The basic b-ball pattern with h hands is a same-hand fountain, next-hand cascade, or previous-hand cascade if and only if

$$b \bmod h = 0, 1, h - 1,$$

respectively.

In the following, we list some further more or less obvious consequences of these considerations that nevertheless are worth spelling out individually.

Fix the number h of hands. Then, we can define a sequence whose bth element is a c or an f depending on whether the basic b-ball juggling sequence with h hands is a cascade or fountain, respectively. For example, for two hands we get the sequence $c, f, c, f, c, f, c, f, \ldots$, and for six hands we arrive at the sequence $c, f, f, f, c, f, c, f, f, f, c, \ldots$. Since the type (cascade or fountain) of a basic pattern only depends on the sets of prime factors of b and h, this sequence is periodic and the period is a divisor of the number of hands.

Two hand numbers h_1 and h_2 yield the same sequences if and only if they have the same prime factors. This means that a multihanded alien that has to deal with the same sequence of cascades and fountains as we 2-handed humans has a number of hands that is a power of 2.

Now, consider the following scenario. An h-handed alien picks a random number of balls b and starts juggling the basic b-ball juggling pattern with all its h hands. What is the probability that the alien is juggling a cascade? From what we just said, it is clear that this probability is

$$c(h) = \frac{\phi(h)}{h},$$

where $\phi(1) = 1$ and $\phi(h)$, with $h > 1$, is the number of those among the integers from 1 to $h - 1$ that are relatively prime to h. We remark that ϕ, considered as a function of h, is an important number-theoretic function called the *Euler function*; see also the footnote on page 52. As expected, in the case of two hands, this gives a probability of 0.5.

It is common for people to remark that a 4-ball fountain is not really juggling four balls but just two copies of two-in-one-hand juggled side by side. A remark like this can be very frustrating to someone who has just mastered the basic 4-ball trick. However, it is true that juggling the 4-ball fountain just amounts to playing two copies of the basic 2-ball trick side by side. With cascades, there can be no such objections and, in fact, most jugglers prefer cascades to fountains for this and a number of other reasons. So, maybe it would be more desirable to have a different number of hands to be able to juggle "more" of these perfect patterns. If h is a prime number, for example, then $\phi(h) = h - 1$. This means that we can get the probability of juggling a cascade as close to 1 as we want by choosing a larger and larger prime h. For example, for $h = 7$, the probability is already about 0.87. We can also force the probability to be arbitrarily small. For example, if $p(n)$ is the product of the first n prime numbers, then it is not difficult to show that $c(p(n))$ tends to zero as n goes to infinity.

If you spot two jugglers passing an odd number b of balls, chances are that they are juggling the basic b-ball juggling pattern with four hands. It may come as a surprise that apart from these 4-hand and the basic 2- and 1-hand patterns, hardly any other basic patterns are performed by groups of jugglers. The main reason for this is that either the patterns are too trivial, as in the case of the domino patterns, or the timing is too difficult.

4.9   Juggling Balls and Hands

In the case of a uniform juggling pattern, we noticed that if we let the balls play the role of the hands and vice versa, we arrive at another uniform juggling pattern. This suggests trying to do the same in the case of simple juggling sequences.

Consider our favorite juggling sequence, 441, and let's assume we are juggling this sequence with two hands in the usual manner. Remember that 441 is a 3-ball juggling sequence. Let's label the three balls 0, 1, and 2 and the hands 0 and 1. Now, consider Figure 4.7.

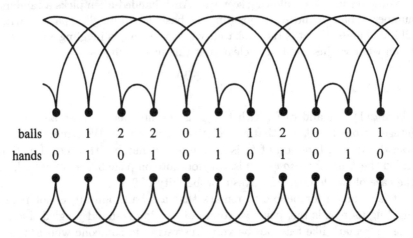

| balls | 0 | 1 | 2 | 2 | 0 | 1 | 1 | 2 | 0 | 0 | 1 |
| hands | 0 | 1 | 0 | 1 | 0 | 1 | 0 | 1 | 0 | 1 | 0 |

FIGURE 4.7. Juggling the juggling sequences 441 and 2 in terms of balls and hands, respectively.

The upper part is the juggling diagram for the juggling sequence 441. Under every beat, we note which hand throws which ball. This gives two periodic sequences. Now, it is clear that every periodic (two-tailed) sequence of integers also gives rise to a juggling sequence. We illustrate this in the case of the sequence that corresponds to the hands and draw the corresponding juggling diagram upside down: Just label the beat points (in order) by the elements of the sequence. Given any beat point, connect it by an arc to the next beat point to its right that carries that same label. In our special case, we end up with the juggling diagram of the 2-ball juggling sequence 2.

So, what we are really doing when we are performing any simple juggling sequence in the usual manner is to juggle this sequence in terms of balls and the sequence 2 in terms of hands. If we were to turn the preceding diagram upside down, we would end up with two balls being juggled by three hands, where the two balls move according to the juggling sequence 2 and the hands to the juggling sequence 441.

It should be clear that we can combine any two simple juggling sequences in such a way that the first sequence is juggled by the balls and the second sequence is juggled by the hands. Let's call the resulting combination a *b-ball h-hand simple juggling sequence*. The only thing we have to worry about is the distribution of the 0s in both sequences. If there are no 0s, then we can combine both sequences in whichever way we want. If we have a 0 in the second sequence (the hand sequence) but no 0 in the first sequence, then we are in trouble since at some point in time there will be a ball landing but no hand ready to catch it. It should be clear how to check whether two given sequences are compatible.

Of course, actually juggling hands and balls is usually quite hard and takes some getting used to. Here are some remarks to get you started if you actually want to give this a try. From the considerations above, it is clear that the 2-ball juggling sequences without 0s are the most useful juggling sequences to be practiced when juggling hands. Prime examples of such juggling sequences are the ground-state sequences

$$2, 31, 411, 5111, 61111, \ldots.$$

In fact, any other 2-ball juggling sequence without 0s can be pieced together from these sequences (up to cyclic shifts). For example, the 2-ball juggling sequence 312231411 splits into 2 (twice), 31 (twice), and 411. Try juggling the juggling sequence 3 with balls while juggling 5111 with your hands. The hand sequence for 5111 is just

$$\cdots 0000111100001111 \cdots.$$

This means that what you want to do is juggle three in one hand for four tosses, then three in the other hand for four tosses, and so forth. Of course, this is quite hard and gets even harder once you try other more complex combinations. Also, really interesting combinations are only possible if we use more than two hands.

4.10   Juggling Labeled Balls

The Flying Karamazov Brothers are a group of famous jugglers. In their show, they wear different colored clothes and start every club-passing routine with sets of clubs of matching colors. They often repeat their passing patterns just the right number of times such that each of the brothers ends up with the clubs he started with, giving a very nice visual effect. Starting with this observation, Arthur Lewbel showed in [72] that this color-matching trick is always possible, no matter which simple periodic juggling pattern you want to juggle.

To make this statement more precise, let us consider the following scenario. We first juggle a b-ball simple juggling matrix of period p. Then, juggling this matrix is periodic with period p in that every p beats the same hands perform the same throws. Now, let's juggle the same simple juggling matrix with balls labeled from 1 to b. We want to show that there is an integer P such that every P beats the same hands perform the same throws with the same balls. This just means that even if we take labelings of balls into consideration, juggling the juggling matrix stays periodic.

To prove this result, we use a form of generalized juggling state. Suppose we are in between two beats. Then, we note the juggling state we are in as usual, except that instead of the 1s in the state we use the labels of the balls. For example, suppose we are juggling a 3-ball simple juggling sequence of maximum throw height 5. Further, suppose that one beat from now ball 2 will land, two beats from now no ball will land, three beats from now ball 1 will land, and four beats from now ball 3 will land. Then, the corresponding juggling state is 20130. The generalized juggling state associated with the moment in time under consideration consists of this juggling state plus the index of the column in the juggling matrix responsible for the throws on the next beat. Clearly, given one of these generalized juggling states, it is possible, using the juggling matrix, to figure out exactly what will happen on any future beat and to reconstruct what has happened on any past beat. Since there are only finitely many generalized juggling states, this implies that there must be two times, some P beats apart, whose corresponding generalized juggling states are the same. Since the relative futures and pasts of these two juggling states are indistinguishable, we conclude that even if we take labelings of balls into consideration, juggling the juggling matrix we started with is periodic.

The color-matching trick for arbitrary simple periodic juggling patterns is a corollary of this general result. The reason for excluding proper multiplex patterns is that whenever a multiplex throw involving at least two balls and throws of different heights or to different hands is made, we have a choice of which of the two balls will perform which throw. Using this choice, it is possible to juggle the pattern with labeled balls in a nonperiodic fashion.

Most of the results in this book relate to juggling scenarios in which the balls are not labeled. In general, it would also be interesting to prove the "labeled" counterparts of these results.

4.11   Decomposing Simple Juggling Sequences

In Section 2.8, we showed that any simple juggling sequence corresponds to a closed loop in some state graph. We observed that the natural decomposition of this loop into smaller loops translates back into a decomposition of the juggling sequence into smaller juggling sequences.

In the following, we want to look at decompositions of juggling sequences in terms of orbits of balls. The general setup is similar to the one in Section 4.9 in that we juggle using h hands (denoted 0 to $h-1$) such that the hands take turns throwing the balls in cyclical order.

Let's first consider the 3-ball juggling sequence 4413 juggled with two hands. Figure 4.8 shows the front view and the juggling diagram of 4413.

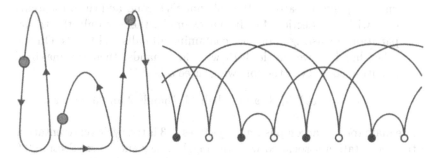

FIGURE 4.8. Front view and juggling diagram of 4413 juggled with two hands.

These two diagrams highlight a decomposition into essentially different ball orbits, which is very useful to keep in mind when we are trying to learn or just describe the corresponding pattern. Ball 0 never leaves the left hand (hand 0) and is always thrown to height four. Ball 1 never leaves the right hand (hand 1) and is always thrown to height four. Ball 3 moves on a circular 31 path between the left and right hands.

In general, given a b-ball juggling sequence s of period p juggled with h hands, we want to set up procedures that allow us to do the following:

1. Describe the essentially different orbits that the juggling diagram of s consists of and enumerate how many there are of each type. Here, two orbits are considered to be essentially different if they are

not the same up to translation in the horizontal direction. In our example, there are two essentially different orbits. One corresponds to the orbits of balls 0 and 1 and the other to the orbit of ball 2.

2. Describe for each type of orbit the order in which throws in s are fitted together to produce the orbit. In our example, this is obvious.

3. List the juggling sequences that correspond to all b orbits in the diagram. In our example, these sequences are 4000, 0400, and 0031.

4. For each ball, figure out which hands it comes in contact with. In our example, ball 0 only touches hand 0, ball 1 only touches hand 1, and ball 2 touches both hands.

5. For each hand, figure out which balls come in contact with it.

6. Construct the *maximal partition* of the balls and hands associated with s (and h). Here, a *partition* is a partition of the balls and hands into sets consisting of at least one ball and one hand each such that the hands in a set are exactly the hands that touch the balls in the set and vice versa. A partition is maximal if it contains a maximal number of sets among all partitions. It is clear that a maximal partition is uniquely determined and that it is an analog to the decomposition of uniform juggling patterns that Shannon's second and third theorems deal with; see Section 4.6. In our example, there is only the trivial partition consisting of one set containing all balls and hands. On the other hand, if we juggle 4413 with four hands, then the maximal partition consists of the following three sets:

{hand 0, ball 0}, {hand 1, ball 1}, {hands 2 and 3, ball 2}.

The main tool for accomplishing objectives 1–3 is the cycle representation of the permutation associated with the juggling sequence s; see Section 2.6 for the definition of the permutation associated with a juggling sequence.

We describe what exactly needs to be done using the sufficiently complex example

$$s = 756970722.$$

This is a 5-ball juggling sequence of period 9. The permutation that corresponds to s is

$$0 \mapsto 7, 1 \mapsto 6, 2 \mapsto 8, 3 \mapsto 3, 4 \mapsto 2, 5 \mapsto 5, 6 \mapsto 4, 7 \mapsto 0, 8 \mapsto 1.$$

The cycle representation of this permutation is

$$(3)(5)(07)(16428).$$

Table 4.2 summarizes the information that can be extracted from this cycle representation.

cycle	sequence	balls/orbits	order
(3)	000900000	1	9
(5)	000000000	0	-
(07)	700000020	1	7-2
(16428)	056070702	3	5-7-7-6-2

TABLE 4.2. Decomposing the juggling sequence $s = 756970722$.

First, every single one of the orbits in the juggling diagram can be described by a juggling sequence. Every such sequence corresponds to one of the cycles as follows. Starting with the juggling sequence of period p consisting only of 0s, replace the ith 0 by the ith element of s if i is contained in the cycle. Note that s has a 0th element. For example, (16428) corresponds to the juggling sequence 056070702. The average of this sequence is the number of balls whose orbits are described by this juggling sequence. Given one such orbit, a second one is just the first one translated a distance p to the right, the third orbit is the second orbit translated a distance p to the right, and so forth until we are back at the first orbit. Note that the cycle (5) corresponds to the 0-throw and is of no importance to what follows. The cycle (3) corresponds to the 9-throw. In general, the elements in s that are multiples of the period correspond exactly to the single-element cycles in the cycle representation.

The (cyclical) order in which throws are made within an orbit can also be extracted from the corresponding cycle—just replace element i of the cycle by the ith element of s. For example, the cycle (16428) turns into (57752), which means that the cyclical order is 5-7-7-5-2.

One more thing to check is whether any of the juggling sequences that correspond to the cycles can be cyclically shifted into any of the others. This is the case if and only if the corresponding two orbits are horizontal translates of each other; that is, they are essentially the same. In our example, this does not happen, which means that there are essentially three different orbits. This takes care of objectives 1, 2, and 3.

To be able to address the remaining objectives, we have to line things up. We do this by requiring that on beat 0 it is the turn of hand 0 to catch and throw. The first ball that gets caught on or after beat 0 is called ball 0, the first ball different from ball 0 that is caught after that ball 1, and so forth. The different orbits are named after the balls they represent. The first element of s is performed on beat 0. To also be able to work with our example, let's say we want to juggle it using six hands.

The first three columns of Table 4.3 summarize how things are lined up in our example; that is, ball 0 is thrown on beat 0, and its orbit is determined by the juggling sequence 700000020, and so forth. The last column is what we are after.

Let's figure out which hands touch ball 0. The sum of the elements in the corresponding sequence is 9. In general, because of the average theorem,

ball	beat	order	hands
0	0	7-2	0,1,3,4
1	1	5-7-7-6-2	0,1,2,3,4,5
2	2	6-2-5-7-7	0,1,2,3,4,5
3	3	9	0,3
4	4	7-6-2-5-7	0,1,2,3,4,5

TABLE 4.3. Decomposing the juggling sequence $s = 756970722$ further.

this sum will be a multiple of the period. Then, hand n catches ball 0 at some point in time if and only if one of the equations

$$(9x + 0) \bmod 6 = n$$

or

$$(9x + 7) \bmod 6 = n$$

has a solution. The first equation has solutions if and only if $n = 0, 3$ and the second has solutions if and only if $n = 1, 4$. This means that ball 0 gets thrown by hands 0, 1, 3, and 4 only.

In the general case, a ball gets thrown for the first time on beat n_{beat}. The first time it is thrown, it is thrown as a t_1-throw, the next time as a t_2-throw, and so forth up to t_k, when things start to repeat. The throws and their order are given (up to cyclic shifts) by the respective entry in column three in Table 4.3. We also set $t_0 = 0$ and $n_{sum} = \sum_1^k t_i$. Note that n_{sum} is a multiple of the period. Hand n touches the ball under consideration if and only if one of the equations

$$(n_{beat} + t_0 + t_1 + \cdots t_{i-1} + n_{sum}x) \bmod h = n,$$

with $i = 1, 2, \ldots, k$, has a solution. In the case of ball 4 in our example, we have $n_{beat} = 4$, $n_{sum} = 27$, $k = 5$, $t_1 = 7$, $t_2 = 6$, $t_3 = 2$, $t_4 = 5$, and $t_5 = 7$.

This takes care of objective 4. The remaining two tasks set in objectives 5 and 6 can be easily accomplished by reshuffling the information that we derived in objective 4. For example, it is clear that hand 0 touches all balls, hand 1 touches all balls except ball 3, and so forth. Finally, the maximum partition of the balls and hands in our example consists of one set containing all hands and balls.

3 balls in one hand

4 clubs

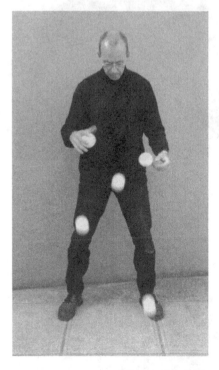

5 bouncing balls

7 balls

9-ball cheat

juggling a braid

inverted pendulum

5
Practical Juggling

Only a very small number of juggling sequences that can be juggled in theory are perceived to be attractive or within the reach of a human juggler. Mind you, the number of attractive and doable juggling sequences is still enormous. In the following, we list and describe a number of the juggling sequences that are particularly attractive from a purely visual point of view or useful for teaching purposes.

Following this, we discuss ways to make juggling easier by taking away gravity in various ways or bouncing balls, and we give descriptions of how these simple ways of juggling have been used in constructing juggling robots.

In the next section, we model the 2-hand juggling of cascades and fountains to deduce how high and how accurate these patterns have to be juggled. We derive a simple model for club throws, which we then use, in conjunction with previous results, to explain various lining-up effects in juggling 2-hand fountains and cascades.

Finally, we summarize the main advantages of the mathematical language for juggling patterns developed in the previous chapters.

5.1 Jugglable Juggling Sequences

Unless otherwise noted, all of the following juggling sequences are supposed to be juggled using two hands in the usual manner. Also, the emphasis is on simple juggling sequences, although the most basic multiplex juggling sequences have also been included.

At this point, it is worth emphasizing again that hardly anything can beat the freely available computer juggling animators when it comes to visualizing one of these tricks and breaking it into manageable pieces. Before you read on, I again recommend that you download one of these programs (described in Section 1.3) and use it to visualize the patterns in this section.

The simplest juggling sequence is 0—to juggle it, you just stand there and do nothing. The next simplest sequence is 1—you zap one ball back and forth between your left and right hands. Next is 2—you again just stand there with one ball held in your left hand and the other ball in your right hand. All these "tricks" should not be missing from any juggling and math lecture and can actually be "performed" in a very funny manner.

In fact, things really start happening with two balls, and I highly recommend including some 2-ball juggling sequences in any demonstration of mathematical juggling. What makes them ideal for demonstrations is that it is really easy to see what exactly is happening, most important features of general juggling sequences are already present in the 2-ball juggling sequences, they are easy to learn and to perform flawlessly, and so forth.

My favorite 2-ball sequences are 2, 31, 312, 330, 411, 40, and 501; see Figure 5.1 for front views of the respective basic 2-ball patterns.

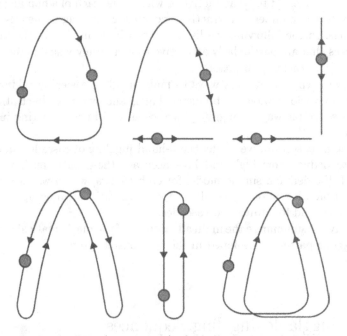

FIGURE 5.1. Front views (not to scale) of 2-ball patterns whose corresponding juggling sequences are 31, 312, 411, 330, 40, and 501. In 501, the two balls move along two disjoint circles.

- 31, in a sense, is the simplest nontrivial juggling sequence; see the first diagram in Figure 5.1 for a front view. It is also a sequence that many people can juggle. The two balls are moving in a circle. One of the hands always performs 3-throws, while the other hand always performs 1-throws. This is an asymmetric pattern. Depending on which hand makes which toss, the balls move in a clockwise or counterclockwise direction around the circle.

- 312 is playing 31 in one direction for two beats, then you pause and play 31 in the opposite direction for two beats, then you pause again, and so forth. This pattern shows one of those strange 2-throws/holds in action; see the second diagram in Figure 5.1 for a front view.

- 330 is the 3-ball cascade with one ball missing; see the fourth diagram in Figure 5.1 for a front view. If you want to demonstrate it, juggle the 3-ball cascade, drop one ball, and keep on juggling as if nothing has happened. One period of this pattern just amounts to swapping the two balls between the hands. Juggling 330 first for one period, then for two, three, and four periods, and so on, is exactly what novice jugglers are advised to practice before trying the 3-ball cascade. Most people find that it is the pattern 31 that at first gets in the way of mastering 330 and, eventually, the 3-ball cascade. What happens is that they execute the first 3-throw with the right hand, but then, instead of making a controlled 3-throw from the left, they panic and try to get rid of the second ball as quickly as possible. This invariably results in a 1-throw.

- 411 is the third element in the sequence of 2-ball juggling sequences

$$2, 31, 411, 5111, 61111, 711111, \ldots;$$

see the third diagram in Figure 5.1 for a front view of 411. You can use this sequence of sequences to check what the maximum height of throws is that you can perform. The aim is to throw one ball as high as you can and then zap the second ball as many times as possible between the left and right hands. Note that the number of zaps plus two is the height of the high throw.

- 40 is basically two balls in one hand while the other hand does nothing; see the fifth diagram in Figure 5.1 for a front view. In fact, suppose you see someone perform this trick and you do not know the underlying beat. Then, you cannot be sure which juggling sequence is being performed. This is because jugglers usually do not move their hands if they don't have to, and therefore the hands usually don't do anything if there is a 0-throw coming up. This means that any of the sequences

$$2, 40, 600, 8000, \ldots$$

(with one hand) is possible. Of course, something similar is true for any other trick. For example, the sequence 3 is indistinguishable from 900 if you don't specify the beat in some way.

- 501 is a nontrivial little gem of a juggling sequence; see the sixth diagram in Figure 5.1 for a front view. One ball will only peak on one side of the pattern while the other always peaks on the other side of the pattern—very striking if you use balls of different colors. Since this pattern includes one 5-throw, it is also a nice first step on the long journey towards a solid 5-ball cascade. Another good practice trick for the 5-ball cascade is the 2-ball sequence 50500.

Once you start juggling 3-ball patterns, people will start taking you seriously as a juggler. Also, most good 3-ball patterns are still fairly easy to learn and to perform well. My favorite 3-ball juggling sequences are 3, 423, 4413, 441, 51, 50505, 52512, and 7131. See Figure 5.2 for front views of some of the respective basic juggling patterns.

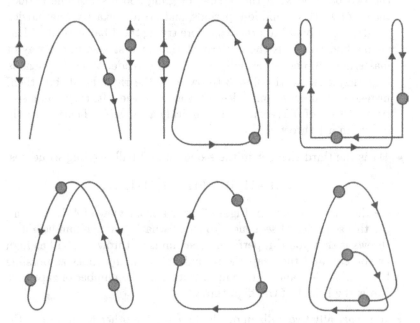

FIGURE 5.2. Front views (not to scale) of 3-ball patterns whose corresponding juggling sequences are 423, 4413, 441, 50505, 51, and 7131.

- 423 is probably the easiest 3-ball juggling sequence; see the first diagram in Figure 5.2 for a front view. One ball each is going up and down on the left and right (4-throws) while one ball is hopping back and forth between the left and right hands (3-throws).

- 4413 is my all-time favorite trick for explaining juggling sequences; see the second diagram in Figure 5.2. This sequence incorporates throws to three different heights that people can "understand," and its structure is very obvious even to nonjugglers. As in the previous trick, there are two balls that move up and down, one on the left and one on the right. The third ball is constantly moving in a circle between left and right. To be able to easily refer to the different balls and throws, I usually make sure that the balls on the outside are of one color while the ball in the middle is of a different color.

- 441 is also very visual and has the overall appearance of a box; see the third diagram of Figure 5.2.

- 531 is 501 with an extra ball hopping from left to right. It is the first in the nice sequence of juggling sequences

$$531, 73131, 9313131, \ldots ,$$

which is very similar to the sequence of 2-ball juggling sequences that we considered above.

- 50505 is also known as the *snake*; see the fourth diagram in Figure 5.2. You start with three balls in your dominant hand and throw all three in rapid succession such that the last ball is in the air before the first one lands. The overall appearance is that of the three balls "snaking" back and forth between the two hands. This is the 5-ball cascade with two balls missing. The second way of playing the 5-ball cascade with two balls missing is 55500, the 3-ball *flash*.

- 51 is the 3-ball shower; that is, all balls move in a circle. Many non-jugglers are under the impression that this is the basic 3-ball juggling pattern. It is not and is actually a lot harder than the 3-ball cascade; see the fifth diagram in Figure 5.2 for a front view of 51.

- 52512 is also known as *Allen Knutson's baby juggling pattern*. The baby is held in the left hand, the right hand executes two 5-throws, the baby is passed to the right hand, the left hand catches the two incoming balls and sends them on their way again as 5-throws, the baby is passed back to the left hand, and so forth. Note that the baby acts as the third juggling ball.

- 60 is *three in one hand* or half of a 6-ball fountain. If you are able to juggle this pattern, it is also a good idea to learn the 3-ball cascade using just one hand. This means that the hand plays the roles of both the left and the right hands. This is a good trick to know if you want to demonstrate that any alternating juggling pattern can be juggled with just one hand and at the same time impress your audience.

- 7131 looks like the last diagram in Figure 5.2.

With four balls, things get confusing for most spectators. Attractive 4-ball juggling sequences whose structure can still be made fairly transparent include 4, [33]33, 53, 534, 552, 5551, 633, and 71. Also check out Probert's book [94], which is the best source for good 4-ball tricks. In addition, it contains lists of attractive 5-, 6-, and 7-ball tricks.

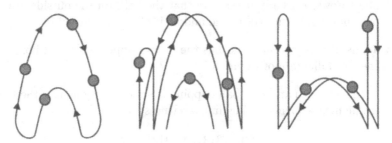

FIGURE 5.3. Front views (not to scale) of 4-ball patterns whose corresponding juggling sequences are 53, 534, and 633.

- [33]33 is the easiest way of juggling four balls. Basically, this is the 3-ball cascade in which one ball gets "doubled up." Similarly, [33][33]3 and [33] are the easiest ways of juggling five and six balls. In fact, multiplex "cheats" are very good tricks to know when it comes to silencing the heckler who is always asking for "one more ball." Provided you have not demonstrated any multiplex juggling before, one very funny and surprising way of finishing off a ball routine is first to go to great length to explain exactly how difficult it is to fit nine balls in your hands, let alone juggle nine balls, and then to just juggle nine balls as a juggling sequence [333]. This can be done with nine small balls, or, what is even funnier and absolutely foolproof, with three triangles that have been glued together from three tennis balls each. It is surprising how long it usually takes people to realize that you are really cheating here. Anyway, audiences love this one.

- 53 is the 4-ball *half-shower*, a very useful and quite slow pattern; see the first diagram in Figure 5.3. However, it is not so easy to get the timing right. Many jugglers end up throwing simultaneously with both hands when trying to juggle this sequence. The sequence 53 is also the first in the following sequence of juggling sequences:

$$53, 7333, 933333, \ldots .$$

This sequence is very similar to the sequences of juggling sequences that we considered for two and three balls.

- 552 is a fairly easy pattern in which every ball touches every hand. It is a good trick to play immediately after one of your spectators has complained that the 4-ball fountain is just juggling two in one hand with both hands. It has the additional redeeming feature of looking roughly like a cascade.

- 534 is arguably the most attractive 4-ball juggling sequence. It is important to maintain the height differences between the throws to different heights; see the middle diagram in Figure 5.3.

- 5551 is easy to understand but fairly hard to juggle well.

- 633 is closely related to 441; see the last diagram in Figure 5.3.

- 71 is the 4-ball shower.

You are wasting your time if you are trying to master complex 5-ball juggling sequences as long as your 5-ball cascade is not very solid. Try juggling 64, which splits into three in one hand juggled with one hand and two in one hand juggled with the other hand. Also worth a go is 66661, which is very similar to the 3-ball pattern 441, and the 5-ball shower 91. Apart from this, it is best to choose and pick among the juggling sequences that are packaged with the various juggling animators and the ones listed in [94].

There are also a number of very doable 5-ball multiplex juggling sequences. We already mentioned [33][33]3, the 3-ball cascade with two of its balls "doubled up." Also, [32] is worth a go. In fact, [32] can be juggled in essentially two (or more, if you are picky) different ways, depending on how you choose to interpret the 2-throws. On the one hand, you can just continuously hold one ball each in your left and right hands while at the same time juggling the 3-ball cascade. On the other hand, you can actually perform the 2-throws.

5.2   Juggling Made Easy

In [28], page 85, Charlie Dancey remarked jokingly, "As you juggle a 3-ball cascade, be aware of the *fourth* ball that makes this wonderful trickery possible—it's the one that you are standing on." Of course, he is right; after all, gravity is what makes toss juggling as we know it possible and as difficult as it is. There are tricks to make juggling easier—that is, preserve the way the balls are moving around one another while reducing the effort, accuracy, acceleration, and so forth needed to keep them going. In the following, we will describe some of these tricks.

5.2.1 Zero-Gravity Juggling

The first thing that comes to mind is to play with gravity. What if we could reduce gravity or even switch it off completely? Although Neil Armstrong missed out on the opportunity to try juggling on the Moon, on April 15, 1985, astronaut Don Williams made juggling history by juggling an assortment of apples and oranges in space at zero gravity; see [44] for a detailed report. Given that an object continues at a constant speed on a straight path after being tossed, he first juggled a 3-fruit pattern by tossing the fruits from hand to hand across his body. Commenting on his experience, Williams said, "You have to toss objects much more easily, on the order of 1/4 as hard, to keep them under control. Quick hand movement is not at all an asset as on Earth. As a matter of fact, the slower I moved the better off I was." Williams also tried a four-object shower from left to right and found that it was very important to impart the same velocity on his juggling props to prevent them from colliding with one another.

A good approximation of zero-gravity juggling attempted by Williams is possible here on Earth—on a billiard table. When you are rolling a ball on a billiard table, gravity does not play much of a role. Also, almost all juggling patterns are juggled in a plane, so putting this plane in a horizontal position is quite a natural thing to try. As in space, balls will travel in straight lines. With a little bit of practice, it is possible to juggle the patterns that Williams tried. Also, the 3-billiard-ball cascade is possible by rolling the balls along two crossing "runways" on the billiard table in front of you; see Figure 5.4 on the left for a top view of what is happening.

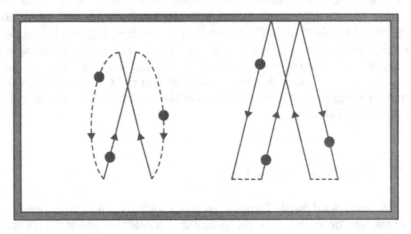

FIGURE 5.4. Simulating two ways of juggling at zero gravity (on a billiard table).

The solid lines indicate where the balls are rolled and the dashed lines where the balls get carried or at least dragged by the hands (left hand on the left, right hand on the right). Even a 5-ball cascade can be juggled in this way, but things get very exhausting and difficult.

However, it turns out that there is a much better way of juggling on a billiard table. What you do is to stand leaning against one of the longer sides of the table and then bounce balls off the opposite side. Figure 5.4 on the right shows the top view of a 3-billiard-ball cascade juggled in this way. Juggling like this is so easy that, even if you are not a juggler, you should be able to juggle three on the first try and five or even seven billiard balls within an hour. I have found that with a little bit of practice even very long runs with nine are not a problem and reasonable runs with eleven are within reach. Passing improbable numbers between two jugglers as well as all kinds of complicated patterns are also possible and a lot of fun. In fact, billiard ball juggling may well be the easiest form of "real" juggling, and it is surprising that hardly anybody seems to know about it.

We only remark that by tilting the billiard table it is possible to simulate, for example, juggling on the Moon. Also, due to the almost complete absence of friction, air hockey tables present two-dimensional juggling environments that are even closer to space juggling than billiard tables. However, air hockey pucks are not as easy to manipulate as billiard balls.

Here is another idea to make things even easier; see also [118], page 864. Build a billiard table in the form of an ellipse such that the focal points of the ellipse are about the width of your shoulders apart. Then, launching a ball in any direction starting at one of the focal points ensures that it will end up at the other focal point. Because of well-known properties of ellipses, the distance that a ball travels between the two focal points is independent of the direction in which it moves initially (unless you roll the ball straight across). This means that when we juggle billiard balls on such a special table by launching and picking them up at the focal points, we only have to worry about keeping the initial velocity at which we launch the balls constant, get the spacing between the balls right, and stir the balls into orbits that prevent them from colliding with one another.

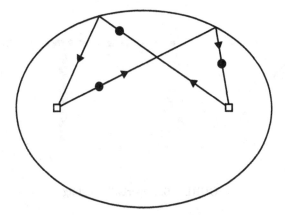

FIGURE 5.5. Juggling on a billiard table shaped like an ellipse with hands catching and throwing at the focal points.

5.2.2 Bounce Juggling

What we have been advocating above is *bounce juggling* on billiard tables; that is, horizontal bounce juggling. Vertical bounce juggling is also possible and is, in fact, quite popular. To practice bounce juggling, you need a flat hard floor and bouncy balls such as lacrosse balls or high-quality silicon balls. Balls are juggled by bouncing them off the floor. Bounce juggling is much easier than toss juggling and is almost effortless. Learning to solidly bounce juggle five balls is a matter of days or weeks instead of months or years as with 5-ball toss juggling. In toss juggling, you have to put a lot of energy into throwing a ball. On the other hand, when catching the ball, it is traveling at maximum velocity and you have to absorb the same amount of energy that you first put into throwing the ball before you can throw it again. In bounce juggling, you have to spend virtually no energy when you launch the balls, and at the moment when you catch the balls, they travel as slowly as possible, and you hardly have to absorb any energy before you can send them on their way again.

In an obituary for Claude Shannon [79], Arthur Lewbel reported that Shannon, the famous information theorist, once asked him, "Do you mind if I hang you upside down by your legs?" Shannon, who had been very interested in juggling, unicycling, and so forth had realized that although bounce juggling is much easier than toss juggling, throwing upwards is physiologically easier. His idea was to combine the two by having an expert juggler bounce juggle while he was hanging upside down. Sadly, Lewbel does not go on to tell whether they actually tried this experiment. However, Bobby May, one of the most famous jugglers of this past century, was known to bounce juggle while standing on his head, but this feat is better listed in a section entitled "Juggling Made Hard."

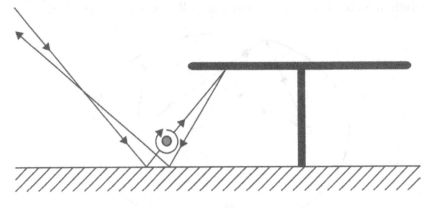

FIGURE 5.6. Superball juggling.

In [118], page 86, Shannon suggests two more related ways of semiautomatic juggling. The first involves *superballs*. If you throw a bounce ball with a grippy surface at an angle against the floor, a considerable fraction of the horizontal component of energy is turned into rotational energy. If it hits the floor a second time, the second bounce will therefore appear very flat. On the other hand, if the second surface it hits is the underside of a table, as in Figure 5.6, the rotational energy will make it return to you along almost exactly the same path, like a boomerang. In fact, you can throw the ball quite arbitrarily and it will return to the hand that launched it. In [65], page 41, the juggler Michael Moschen is shown juggling four balls in this way.

Shannon also suggests bounce juggling in a bowl that is shaped like half of an ellipsoid, with the two focal points contained in the plane that is spanned by the rim of the bowl and the two hands held at the two focal points. This is basically the same idea as juggling in an ellipse except that we are no longer restricted to juggling in a plane. Shannon points out that bouncing balls do not exactly reflect like light rays (or billiard balls), but their behavior is close enough to achieve the desired effect.

5.2.3 Robot Juggling

Shannon was also the first to build a juggling robot or, to be more precise, a bounce-juggling robot. This robot bounce juggles three steel balls on a drum. The inner skeleton of the robot was constructed from an erector set, and on the outside it resembled W.C. Fields, the famous juggler and comedian. Figure 5.7 illustrates how this robot works.

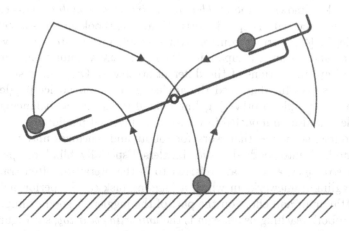

FIGURE 5.7. A simplified rendering of Shannon's juggling robot.

The two arms are fixed relative to each other and, driven by a motor, move by rocking sideways around a fixed axis. The hands are formed as short grooved tracks. A hand catches every time it is in the down position. Then, as the hand moves up, the ball rests against the back of the track.

As the hand rises above the horizontal, the ball starts to accelerate along the track and gets launched when the hand reaches the up position. At the same time, the other hand makes a catch. Christopher Atkeson and Stefan Schaal refined Shannon's original design and built a juggling machine that can bounce juggle five balls; see [107], Section 2.3.

Shannon's robot is just one example of a host of different robots that have been devised to accomplish juggling-related tasks. There are devilsticking robots ([107] 2.4), unicycling robots ([110], [119]), robots that move around by hopping on one leg ([48]), robots that play the Japanese juggling game Kendama ([83]), robots that can balance a broomstick or *inverted pendulum* (see [46]), and so forth.

It proved to be a difficult task to build a robot that can actually toss juggle. However, recently it has been reported that even this has been achieved (see [115]) both by *seeing robots* and *blind robots*. A seeing robot uses sensory input to compute actuary commands for error correction. On the other hand, Shannon's robot is a blind robot in that it does not use active reaction to respond to perturbations and relies exclusively on the geometry of the mechanical device, the kinematics and dynamics of motion, and the properties of materials to stabilize the juggling task execution.

Advice that is given to jugglers trying to master a new trick often proves to be very important in the design of robots that are supposed to perform the same trick. For example, "Watch the top of the club and not your hand (or some other part of the body used for supporting the club)!" is advice that is often given to people trying to master balancing a juggling club. In the theory of control, balancing an object such as a juggling club or broomstick is known as the *problem of the inverted pendulum*. This control problem is discussed in depth in virtually any textbook on this subject. For example, in [46], various control strategies are investigated to solve this problem, and on an accompanying Web page, JAVA animations simulate real-life implementations of the different strategies. Only those strategies are really successful that heed the advice given to our novice juggler. Incidentally, a couple of other juggling-related tasks such as balancing on a unicycle can also be modeled in terms of an inverted pendulum.

Of course, strategies that work for robots and machines may very well translate back into good advice for jugglers. Especially, blind, or *open loop control strategies,* as they are referred to in the literature, often translate into juggling frameworks in which a juggling task can be performed with very little input and effort on the part of the juggler. A good example of a blind "robot" teaching us a lesson is the *unicycling bear toy;* see Figure 5.8. The unicycling bear rides its unicycle perched on a rope, and there is no chance of it falling off because the two weights attached to its balancing pole move its center of gravity below the point at which the wheel of the unicycle touches the rope.

See also [8] and [57] for two more accessible accounts of robot juggling. The references given here are only meant as first contact points with a

FIGURE 5.8. The unicycling bear toy with center of gravity below the tightrope on which it is riding.

very complex field of research. Finally, for more information about Claude Shannon and his contributions to juggling, see [79], [80], [117], and [118] as well as Subsection 7.5.3.

5.3   Real-World Juggling with Gravity and Spin

In [136], Bengt Magnusson and Bruce Tiemann set out to describe some of the basic physical laws that govern the actual juggling of the basic b-ball juggling sequence with two hands in a cascade or fountain pattern; see Figure 5.9. Taking into account the number of balls juggled, gravity, and nonzero dwell time, they deduced some ballpark figures for how accurate and how high balls have to be thrown when juggling these patterns. Also included in this paper is a simple model for juggling spinning clubs, which is then used to explain why very often all the clubs in the air line up in a very striking manner. In the following, we summarize these results. We only remark that this paper also contains the first account of the basics of juggling sequences.

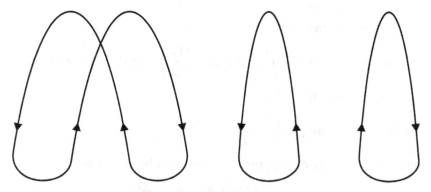

FIGURE 5.9. Juggling a cascade or fountain pattern.

5.3.1 Accuracy and Dwell Time

We first specify again how exactly we are juggling the b-ball cascade for odd b and the b-ball fountain for even b. Catches are made on equally spaced beats, and the two hands take turns throwing the balls. The time between two catches of the same hand is denoted by t. The dwell time of a ball in a hand is a constant $t\theta$, where $0 \leq \theta \leq 1$. This means that the vacant time of a hand is $t(1 - \theta)$ and the flight time of a ball is $bt/2 - \theta t$. This also means that the height of a throw is

$$h = \frac{1}{32}g(b - 2\theta)^2 t^2,$$

where, as usual in physics, $g = 9.81$ m/s^2. For most jugglers, θ tends to be very close to 0.5. The reason for this is that in this case the very strong rhythm of one hand catching a ball at exactly the same time as the other hand is throwing a ball governs the juggling effort; see also Section 7.6 for more details on another line of research dealing with this issue. On the other hand, t has been observed to vary from 0.2 to 0.8 seconds.

How accurately a juggler has to aim every single throw is most easily expressed in terms of the time t between catches of the same hand and the height h of the throws; high throws must be aimed better than low throws, and the time t determines how much time the juggler has to aim this well. It is easier to aim well if you have a lot of time to do so than if you have very little time.

For the moment, let's not worry about possible collisions of balls in the air. In fact, let's assume that the balls are points. Between being thrown and being caught, a ball travels on a parabolic arc of width w (approximately the distance between the hands) with an error margin of Δw on both sides. Balls are thrown with velocity v_0 at an angle of α from the vertical with an error of $\Delta \alpha$, which gives rise to the error Δw.

Keeping in mind that we are throwing to height h, we derive the following identities:

$$v_{0y} = (2gh)^{\frac{1}{2}},$$
$$v_{0x} = v_{0y} \tan \alpha,$$
$$w = tv_{0x} = \left(\frac{8h}{g}\right)^{\frac{1}{2}} v_{0y} \tan \alpha = \left(\frac{8h}{g}\right)^{\frac{1}{2}} (2gh)^{\frac{1}{2}} \tan \alpha = 4h \tan \alpha,$$
$$\alpha = \arctan\left(\frac{w}{4h}\right),$$
$$\alpha + \Delta \alpha = \arctan\left(\frac{w + \Delta w}{4h}\right).$$

Since for small x we can approximate $\arctan(x)$ by x, we have

$$\Delta \alpha \sim \frac{w + \Delta w}{4h} - \frac{w}{4h} = \frac{\Delta w}{4h}.$$

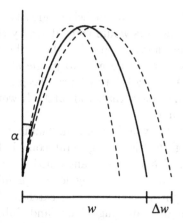

FIGURE 5.10. Variations of the angle α translate into variations of the width w of the parabolic arc on which a balls travels.

The error margin Δw is around 30 centimeters for most jugglers, and the highest patterns juggled by the world's best jugglers with any consistency are about 6 meters high. This means that Sergei Ignatov, who juggles 11 rings at this height, has to aim every ring to within 0.6° just to be able to catch it at the far end of its trajectory. Of course, the aim has to be even better to ensure that there won't be any collisions.

With the approximation above,

$$\frac{\Delta\alpha}{\alpha} \sim \frac{\Delta w}{w};$$

that is, the fractional error in the angle is independent of the throw height. For most jugglers, the width of the parabolic arc w is about 90 centimeters. This means that this fractional error is about $1/3$.

We have seen that the maximum throw height is dependent on $\Delta\alpha$. This means that the number of balls that can be juggled by a juggler is limited by $\Delta\alpha$, the dwell time $t\theta$, and by the time t between consecutive throws with the same hand. The more accurate he can throw, the shorter his dwell time, and the faster he can juggle, the more balls he will be able to juggle. However, for large numbers of balls, a strong rhythm is essential, and we may assume that $\theta = 0.5$. Also, for a "typical juggler," we may assume that t is around half a second; that is, there will be four throws per second. Table 5.1 shows the height of throws in the basic b-ball juggling pattern of a typical juggler here on Earth.

number of balls	5	6	7	8	9	10	11	12	14	15
height of throw	1.2	1.9	2.8	3.8	4.9	6.2	7.7	9.2	12.9	15

TABLE 5.1. Height of the the basic b-ball juggling pattern in meters juggled with four throws per second and catches performed at the same time as throws.

If we were to juggle on the Moon, where gravity is about 1/6 of that on Earth, the different heights would come down by the same factor. This means, for example, that a typical juggler who is able to juggle seven balls here on Earth could expect to be able to juggle about 15 on the Moon. Similarly, people like Anthony Gatto who can keep nine balls in the air for extended periods of time (more than 100 catches!) would be able to juggle about 20 balls on the Moon.

We should point out that world records in "numbers" juggling, just like a juggler's ability to juggle b objects in general, come in two categories. For a juggler to be able to *flash* b objects means that he can consecutively throw and catch all b objects once. For a juggler to *qualify* juggling the same number of objects requires $2b$ throws and catches. The most commonly used props for toss juggling are rings, balls, and clubs. Table 5.2 lists the current world records with these different kinds of props. To achieve these world records, a number of different patterns are used besides the ones we are focusing on in this chapter; for example, simultaneous fountains for even numbers of objects and shower patterns for rings.

	flash	qualify	> 50 catches
rings	12	10	9
balls	12	10	9
clubs	9	7	7

TABLE 5.2. Different kinds of world records; see [81].

The "unofficial" record for flashing rings is 14.

5.3.2 Why Clubs and Balls Line Up

Of course, no matter what props we are really juggling, the analysis above still applies. However, when juggling clubs, we also have to take into account that these objects rotate in flight and have to be caught at their handles. Depending on how many times a club spins, a throw with a club is called a *single spin*, *double spin*, and so forth. Also, juggling *five doubles* means juggling five clubs with double spins, juggling *three singles* means juggling three clubs with single spins, and so forth.

By tossing a club higher, it will spin longer, and it seems natural to expect that a double spin would have to be thrown four times as high as a single spin. This is not the case. When you toss the club higher you also make it spin faster, and a double spin ends up being exactly twice as high as a single spin, a triple spin three times as high as a single spin, and so forth.

To model a club throw, we may assume that the club is thrown with the juggler's lower arm as the pivot arm rotating around the elbow. Let l be the distance between the elbow and the center of mass of the club, and let a be the distance between the elbow and the tip of the handle (that is, basically the length of the lower arm; see Figure 5.11). If a juggler uses a lot of wrist action (this varies from juggler to juggler), the distance l may not be exactly equal to the distance between the elbow and the center of mass of the club. However, there will be an effective lever arm length l_{eff} that stays fairly constant for any individual juggler when the juggler is juggling different numbers of clubs, and this effective arm length can play the role of l in the following analysis.

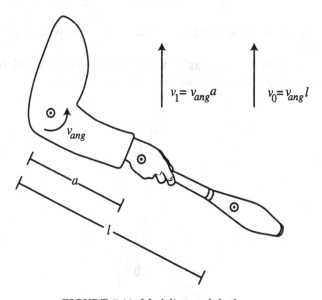

FIGURE 5.11. Modeling a club throw.

The center of mass of the club is released with velocity

$$v_0 = v_{ang}l,$$

where v_{ang} denotes the angular velocity of the pivot arm. Therefore, the club stays in the air for a time

$$t_{air} = \frac{2v_0}{g} = \frac{2v_{ang}l}{g}.$$

On the other hand, the tip of the handle starts with velocity $v_1 = v_{ang}a$. In the center-of-mass frame, the tip of the handle has speed

$$v_0 - v_1 = v_{ang}(l - a).$$

This means that the tip of the handle rotates with the same angular velocity v_{ang} as the pivot arm. Since the handle has to rotate by an angle of $2\pi s$ for s spins, it must stay in the air for a time

$$t_s = \frac{2\pi s}{v_{ang}},$$

which must be equal to t_{air}. We conclude that

$$v_{ang} = \left(\frac{s\pi g}{l}\right)^{\frac{1}{2}}.$$

Therefore, the corresponding throw height is

$$h_s = \frac{1}{8}gt_s^2 = \frac{1}{8}g\left(\frac{2\pi s}{v_{ang}}\right)^2 = \frac{s\pi l}{2}.$$

Note that h_s is linearly dependent on the number of spins s and is independent of g.

In an ideal 5-club cascade juggled with double spins, all the clubs in the air will be lined up perfectly at all times; see Figure 5.12.

FIGURE 5.12. Five clubs juggled in an ideal cascade pattern with double spins are lined up at all times. The numbers on the clubs indicate the order in which they were thrown in the air. Clubs 1 and 2 are falling while the other clubs are rising.

Here is why this happens. In addition to the parameters that we introduced already, let b be the number of objects/clubs juggled. Remember that t denotes the time between two consecutive throws with one hand.

This means that the time between a throw with the left hand and the subsequent throw with the right hand is $t/2$. Therefore, the angular separation between two clubs at the time of release is

$$\gamma = \frac{1}{2}v_{ang}t.$$

With

$$h_s = \frac{1}{2}s\pi l = \frac{1}{32}g(b-2\theta)^2t^2,$$

we get

$$l = \frac{g(b-2\theta)^2t^2}{16s\pi}$$

and

$$v_{ang} = \left(\frac{s\pi g}{l}\right)^{\frac{1}{2}} = \frac{4s\pi}{(b-2\theta)t}.$$

Combining these equations yields

$$\gamma = \frac{1}{2}v_{ang}t = \frac{2s\pi}{b-2\theta}rad = \frac{s\,360°}{b-2\theta}.$$

This means that clubs will be parallel with $\gamma = 180°$ only if $b = 2(s+\theta)$. Since both b and s are integers, $\theta = 0, 0.5$, or 1. Only $\theta = 0.5$ is physically possible, which leaves us with $b = 2s+1$. Since most jugglers juggle with θ very close to 0.5, and since three clubs are usually juggled with single spins, five clubs with double spins, seven clubs with triple spins, and so forth, parallel clubs are a fairly common sight. Parallel clubs are also possible for an angular separation of $\gamma = 360°$; for example, this happens in the case of five clubs juggled with quadruple spins.

Of course, there are only very few jugglers who are able to juggle five clubs with quadruple spins in practice. However, computer juggling simulators such as *JoePass!* (see Section 1.3) allow you to observe perfect jugglers juggling any number of clubs using any number of spins, and it is a pleasure to see all our theory work out perfectly in practice. In fact, Figure 5.12 is based on a screen shot of the juggling animator *JoePass!* in action. Always assuming $\theta = 0.5$, other interesting angles, numbers of balls, and numbers of spin combinations to experiment with are: 90°, five singles (Bruce Tiemann can do this); 120° degrees, four singles or seven doubles; and 270°, five triples.

While you are conducting your virtual (or real) club-juggling experiments, it is best to look at the virtual (or real) juggler from the side. Now, while you are at it, also try the following. Switch to a front view, replace clubs by balls, and let's first assume that our virtual juggler is juggling an odd number of balls. Watch for a while and you will notice a different kind of lining up happening. Follow two balls that are thrown one after the other. When the first ball peaks, the other balls are all lined up horizontally in pairs of two as in Figure 5.13 on the left. This is the first kind of

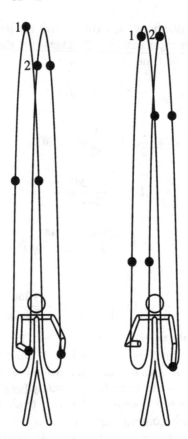

FIGURE 5.13. Seven balls juggled in an ideal cascade pattern line up in two essentially different ways.

horizontal lining up that can happen. Shortly after the first ball peaks, it ends up at the same height as the second ball. At this time, all other balls except for the lowest one are again lined up horizontally in pairs of two as in Figure 5.13 on the right. This is the second essentially different kind of horizontal lining up that can happen. It is easy to see why such linings up of the balls happens, considering the fact that intervals between throws are constant and the fact that it takes a ball the same amount of time to climb from a certain level to the point where it peaks as it takes the ball to fall from this point back to the level under consideration. If we are dealing with an even number of balls, also two essentially different horizontal linings up can be observed. In this case, as the first ball peaks, all the other balls line up in pairs except for the lowest one. If the first and the second line up, all the other balls line up, too.

See the article by Magnusson and Tiemann for photos of the authors juggling clubs with the clubs lining up as predicted. Arthur Lewbel discussed the lining up of balls in the horizontal direction in [76]. Also part of this article are two photos of jugglers juggling five and seven clubs that illustrate the two different types of horizontal linings up in action.

Note that our initial assumption that $\theta = 0.5$ is really a third kind of alignment. After all, what it means is that one hand catches at the same time that the other hand makes a throw; see also the remarks in Section 7.6.

For further information about the physics of juggling, see [63], [64], [66], and the articles by Arthur Lewbel listed in the References.

5.4 What Is All this Numbers Juggling Good for?

We have already given a number of answers to this question in the previous sections. Here, we want to summarize again the answers we have come across so far and complement them with a couple of new ones.

Juggling sequences, or more generally juggling matrices, capture the "mathematical skeleton" of juggling patterns. We can use them to very effectively accomplish a number of tasks that are important to jugglers and mathematicians.

Record and Communicate

In 1984, the magazine *Juggler's World* printed a series of six photographs showing six instances of a juggler performing a complicated trick as an "answer to people who ask why we don't try to teach more tricks via the magazine." It turns out that the trick performed in the pictures was a 4-ball trick that is fairly easy to master once you know that its juggling sequence is 5623. This very nice example illustrating the difficulties of communicating juggling patterns was taken from [68].

Prior to the introduction of juggling sequences, the only really effective way of recording a complicated juggling pattern was by videotaping someone performing it. However, if you have ever tried to reconstruct a trick from a video recording, you will know that even this can be a very frustrating and time-consuming exercise.

No method of recording and communicating can beat juggling sequences in terms of compactness and completeness when it comes to describing patterns made up of simple tosses involving not much in terms of contortions.

Juggling sequences are the right language for teaching computers to juggle. In fact, all pattern generators and animators are built using the theory of juggling sequences. These animators are the perfect front end for finding and decoding juggling sequences and in most cases are also able to completely replace video recordings.

Enumerate and Create New Interesting Patterns

Using algorithms that are based on results in this book, computers have been programmed to enumerate all juggling sequences satisfying any con-

ceivable set of constraints. Many new interesting juggling sequences have
been found in this way. Since we now know "all" possible juggling sequences, what remains to be done is to identify those that, in themselves,
are interesting from either a juggler's or a mathematician's point of view.
Furthermore, we also want to identify all those juggling sequences that can
be turned into interesting tricks by adding the usual or new contortions.

Establish Relationships and "Divide and Conquer"

Clearly, if two juggling patterns share the same underlying juggling sequence, then it makes sense to say that they are very closely related. Having
mastered one should greatly facilitate mastering the other one.

Other relationships between juggling patterns also turn out to be important and can most easily be detected by inspecting the underlying juggling
sequences. For example, a juggling sequence may be the concatenation of
two other juggling sequences, a juggling sequence may be "spliced into" a
second one, as we described it in the section on state graphs, a juggling
sequence may use the same tosses as another juggling sequence although
in a different order, and so forth. Knowing about these different kinds of
relationships will enable you to break down the problem of mastering a
trick into more manageable subproblems.

Also, if you want to find out how you can smoothly move from one pattern
to the next, tools such as state graphs are very helpful. For example, if you
know a trick is a ground-state b-ball trick, then you can go into it from
the basic b-ball cascade or fountain without any transitions. Or you can
concatenate it with another b-ball ground-state trick to arrive at a new
trick. To find the shortest and/or most efficient way to get into an excited-
state trick or from one to another, state graphs or computer programs
based on them are very helpful.

Teach Juggling and Mathematics

By now, it should be clear why juggling numbers are extremely useful in
teaching and learning new juggling tricks.

Since 1985, mathematicians such as Ronald Graham and Colin Wright
have given hundreds of public lectures on juggling sequences in which they
combine a juggling performance with a mathematics lecture. Lectures such
as these are a great way to rectify the public's (dangerous) perception that
mathematics is a hard, boring, and largely useless discipline practiced by
equally boring and useless individuals.

In the math-and-juggling framework, it is very easy to communicate that
mathematics can be the exact opposite of what people think it is: not too
hard, fun, and useful. Also, here are mathematicians who can do the most
amazing tricks, present their ideas in an extremely entertaining manner
and, in general, do not conform to the image of someone straight from the
ivory tower.

Talks about juggling numbers are an ideal way to showcase how mathematicians extract the mathematical skeleton of a natural phenomenon (juggling patterns) by breaking it into its defining and quantifiable pieces (throws) and capturing the rules that make them work together, how they use this skeleton to come to a better and much more complete understanding of this phenomenon (for example, algorithms to generate all juggling tricks), and how they use their mathematical model to evolve the phenomenon beyond its natural limitations (have a virtual juggler juggle 50-ball patterns).

To Go Where Nobody Has Gone Before

This last point may even become more important in the future. Even now, computer-generated movies such as *Final Fantasy* feature virtual actors that are almost indistinguishable from real actors. Who knows, maybe sometime in the not too distant future computer jugglers will choreograph virtual "life-like" juggling performances of similar sophistication in which the performers juggle more objects in more complex patterns than any human will ever be able to do. Any virtual performance will be based on computer juggling animators, which, in turn, are based on juggling sequences.

6
Jingling, or Ringing the Changes

In this chapter, we give an introduction to the ancient art of change ringing, the beautiful mathematics that goes with it, and two new, or at least not very well-known, links between change ringing and the kind of numbers juggling developed in the previous chapters.

Most of the mathematics in this chapter is based on the papers by the mathematician Arthur T. White listed in the References.

6.1 Enter a Band of Ringers

The art, science, and exercise of *change ringing* or *campanology* is all about "ringing" distinguished sequences of permutations of b bells. We start by defining what exactly we mean by this.

6.1.1 Basic Definitions

Given a set, or *ring*, of b bells in a bell tower, let us arrange these bells in descending order of pitch and call them 1, 2, ..., b, respectively. Bells 1 and b are also referred to as the *treble* and the *tenor*, respectively.

In change ringing, bells are rung one bell at a time. In fact, we may assume that the bells are rung to a constant beat, one bell on every beat.

A *change* is the ringing of the b bells in some order; that is, a change is basically the "ringing" of a permutation of $[b]$, the ordered set consisting of the integers 1 to b in their natural order. Of course, this means that there is a total of $b!$ different changes. Ringing the identity permutation/change is called ringing *rounds*.

Ringing the changes means to ring a *b-bell ringing sequence* of changes; that is, a sequence of changes on b bells that has the following properties:

Properties of Ringing Sequences

(B1) The first and last changes rung are both rounds.

(B2) Every change in the sequence apart from the first and last is rung exactly once.

(B3) From one change to the next, no bell moves by more than one position in its order of ringing.

Condition B3 implies that if a bell changes its position from one change to the next, it does so by swapping its position with the bell that precedes it, or the bell that follows it, in the order of ringing.

When change ringers think about ringing sequences, the last change (rounds) is always taken for granted and not spelled out. We will adopt this custom when it comes to recording ringing sequences.

Where do Properties B1, B2, and B3 come from? In the case of B1 and B2, this is not entirely clear. However, it has been ventured (see, for example, [144]) that Property B1 guarantees musicality and Property B2 ensures some sort of perfection. After all, bells are usually rung "to the glory of God," and it only seems right to strive for a similar kind of perfection in the ringing of bells as in the churches that contain them. On the other hand, Property B3 is forced on us by the way the bells are hung and rung; see Figure 6.1.

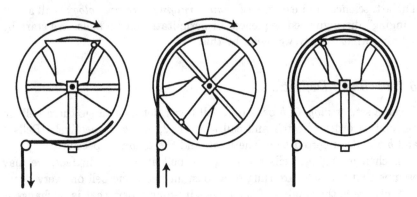

FIGURE 6.1. A bell swings almost full circle. It rings when struck by the clapper on its way up to the balance position.

A bell swings full circle in the clockwise direction starting at the balance position depicted in the left-hand diagram all the way to the balance position depicted in the right-hand diagram. This is followed by a full-circle swing in the counterclockwise direction, which brings the bell back to its starting position. As the bell rises in the course of a swing, the clapper suspended inside the bell strikes the bell edge and a single note sounds. Usually, there is one ringer in charge of every bell who keeps the bell in motion by pulling on the rope whenever the bell reaches a balance position. Note that one end of the rope is attached to a large wheel to which the bell itself is fixed. As the bell swings back and forth, the rope winds around the wheel. With this arrangement, very little force is needed to keep a bell in motion. The way to swap two adjacent bells in a change is to swing the bell that comes first in the change all the way up and hold it there in the balance position until the next bell, which has been given a little less momentum so that it does not quite reach the balance position, is pulled down ahead of the first bell. Balancing a bell is where things get tricky, and it is not possible to move bells with any kind of control by more than one position from one change to the next.

The ultimate aim of b-bell change ringing is to ring an *extent on b bells* or *b-bell extent*; that is, a b-bell ringing sequence containing all possible changes on b bells. Such a maximal ringing sequence is made up of $b! + 1$ changes. Table 6.1 lists extents on one to four bells in the traditional (self-explanatory) notation used by bell ringers. As we already pointed out, we never note the last change of a ringing sequence since we know it coincides with the first change. The 1-bell and the 2-bell extents are unique. The 3-bell extent in the table and its *inverse*—that is, the extent we arrive at by reversing the order of the changes—are the only extents on three bells. The 4-bell extent in the table is an extent of Plain Bob Minimus, the most popular 4-bell extent among change ringers. In the following, we will refer to lists of changes corresponding to ringing sequences as in Table 6.1 as *ringing arrays*.

1	12	123	1234	1342	1423
	21	213	2143	3124	4132
		231	2413	3214	4312
		321	4231	2341	3421
		312	4321	2431	3241
		132	3412	4213	2314
			3142	4123	2134
			1324	1432	1243

TABLE 6.1. Extents on one, two, three, and four bells in the notation used by change ringers.

Traditionally, bell ringers are required to memorize their part in performing a ringing sequence; notes are not allowed in the bell chamber. Considering that especially in the case of extents on a large number of bells, these sequences tend to be very long and can take hours to ring, ringing sequences that are actually performed are based on certain simple algorithms called *methods* and *principles* that are easy to memorize; we will define these terms in our discussion of the mathematics of change ringing in Section 6.4. At this point, we only summarize the algorithm that ringers memorize to be able to reconstruct the 4-bell extent in Table 6.1 on the fly: (1) start with rounds; (2) exchange two pairs of adjacent bells; (3) exchange the middle pair of adjacent bells; (4) keep repeating steps 2 and 3 until after eight changes the order would return to rounds; (5) to avoid this, swap instead the last pair of bells; (6) repeat everything that you have done so far another two times and you are ready to return to rounds on the next change.

Before we describe what change ringing has to do with juggling and focus on the mathematics behind the effective construction of ringing sequences, here are some more remarks about change ringing in general.

6.1.2 History and Practice of Change Ringing

Change ringing started in England in the middle of the seventeenth century. Today, there are about 6000 churches and a few secular buildings that house rings of bells that are used for change ringing. Most of these rings are in English-speaking countries. In fact, there are only about 100 outside the U.K. and Ireland. A ring of bells usually contains 3, 4, 5, 6, 8, 10, or 12 bells. I know of three that contain 16 bells, which seems to be the maximum number of bells that has ever been used for change ringing.

The bells in a ring are tuned in a scale. The heaviest bell used for change ringing is the tenor "Emmanuel" at Liverpool Cathedral, which weighs 4 tons, 2 cwt, 11 lbs. However, very few bells used for change ringing are "great bells"; that is, weighing more than two tons.

Of course, a ring of bells is basically a musical instrument; that is, something on which you want to play tunes. This is possible if you ring the bells by striking them with hammers, as is done in a carillon or when the time is rung out. However, when large bells are rung by swinging them as described above, the gap between two sounds is up to three seconds. This and the mechanical constraints outlined above severely limit the number of tunes you would ever want to play in this way. This is a pity since a bell makes a much nicer sound and carries much further when it is rung by swinging it. It is this dilemma that got people started on devising new games to play on this superb musical instrument whose richest sound cannot be used to play tunes. The most popular such "game" is change ringing (there are others).

bells	name	changes in extent	time required
3	Singles	7	14 seconds
4	Minimus	25	50 seconds
5	Doubles	121	4 minutes
6	Minor	721	24 minutes
7	Triples	5 041	2 hours 48 minutes
8	Major	40 321	22 hours 24 minutes
9	Caters	362 881	8 days 10 hours
10	Royal	3 628 801	84 days
11	Cinques	39 916 801	2 years 194 days
12	Maximus	479 001 601	30 years 138 days
16		20 922 789 888 001	1 326 914 years

TABLE 6.2. Number of changes in extents on up to 16 bells and the time required to ring these extents at 30 changes per minute.

When rounds are rung, all other bells have to ring before the same bell is rung again. This means that if the interval between two ringings of the same bell is two seconds, then this is also the average time needed to complete ringing a change. So, we are talking about 30 changes per minute, which is a good ballpark figure. Table 6.2 summarizes how long it takes to ring extents on up to 16 bells at this speed.

The names in the table are the ones commonly used by bell ringers in naming ringing sequences. The odd-bell names reflect the maximum number of pairs of bells that can be exchanged when moving from one change to the next. The 3- and 4-bell extents in Table 6.1 are extents of Plain Bob Singles and Plain Bob Minimus, respectively. The "Plain Bob" part of the name refers to the simple common algorithm used to construct the extents; see Subsection 6.6.1 for more details on this.

To ring an extent of Major is all that seems to be humanly possible. In recorded history, an extent of Major has been rung on tower bells only once without relays. This took place at the Loughborough Bell Foundry beginning at 6:52 a.m. on 27 July 1963 and ending at 12:50 a.m. on 28 July after 17 hours 58 minutes of continuous ringing. The extent rung was an extent of Plain Bob Major, an 8-bell counterpart of the 4-bell extent in Table 6.1. Even the ringings of extents on seven bells constitute substantial physical and intellectual achievements but are fairly common. Usually, when ringing more than seven bells, the goal for change ringers is to ring a *peal*; that is, a ringing sequence that contains at least 5000 changes.

Change ringing is not a hugely popular pastime. However, there have always been plenty of change ringers, and the craft has never been in danger of dying out. Of course, the expression "ringing the changes" has passed into the vernacular, and change ringing has been popularized to a certain extent by Dorothy L. Sayers' Lord Peter Wimsey murder mystery *The Nine Tailers* [106]. Also, extensive change ringing on very special occasions such as the celebration of the new millennium or, here in Australia, the commemoration of the centenary of the Federation, continue to contribute to its popularization.

It takes a beginner a couple of months to learn to ring his or her bell individually, let alone to ring changes as part of a band of ringers. Quite a bit of the difficulty lies in the fact that, unlike juggling balls, church bells are just not very accessible. However, bell ringers always welcome people interested in their hobby and are more than willing to let you have a go, show you their bells, and answer any questions you may want to ask. For more information about bell ringing, see the many excellent Web sites dedicated to the subject and the books [52], [86], [103], [104], and [105]. A number of computer programs are available that ring the changes. A particularly nice free program is Theo van Soest's WINDOWS program *Method Writer*.

6.2 Juggling the Changes

In this section, we describe two not very well-known links between numbers juggling and change ringing.

6.2.1 Turning Bells into Balls

Start by replacing a ring of b bells by a ring of b (custom-made) *bell balls*. Here, a bell ball is a special ball that makes the sound of a bell when it hits your hand, and a ring of bell balls is a set of bell balls that are tuned in the same way as a ring of bells. Now, it is easy to see that you can ring rounds by just juggling the basic b-ball pattern and throwing the bell balls in the air in their natural order.

What about juggling changes and juggling ringing sequences? No problem (unless you count endless hours of practice as a problem). In fact, any b-bell ringing sequence can be juggled with b (bell) balls. Here is how you go about translating a b-bell ringing sequence into a b-ball juggling sequence. As a concrete example, consider the 3-bell extent in Table 6.1. We construct the juggling diagram of the juggling sequence that corresponds to this ringing sequence; see Figure 6.2.

Start by writing down all changes in the ringing sequence in one row of equally spaced numbers such that every one of the numbers is right below and acts as a label for one of the circles marking the beats in the juggling diagram. Now, ball i gets caught and thrown at precisely the beats labeled with an i. It is now a straightforward exercise to complete the juggling diagram by drawing in the orbits of the different balls.

Let r be a ringing sequence on b bells consisting of $c \geq 2$ changes. Then, the juggling sequence that corresponds to it will be denoted by $juggle(r)$.

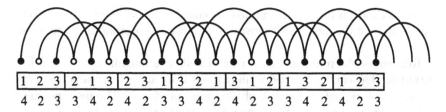

1	2	3	2	1	3	2	3	1	3	2	1	3	1	2	1	3	2	1	2	3

4 2 3 3 4 2 4 2 3 3 4 2 4 2 3 3 4 2 4 2 3

FIGURE 6.2. Translating a ringing sequence into a juggling sequence.

It consists of the first $b(c-1)$ throw numbers that we constructed above. The blocks of numbers in $juggle(r)$ that correspond to the changes in r will be referred to as the *changes of juggle(r)*. This means that $juggle(r)$ consists of $c-1$ changes. The changes of $juggle(r)$ are necessarily ground-state sequences; see Subsection 2.8.2 for a definition of the term ground state. This property follows from the fact that in the course of any change, every one of the balls is caught and thrown exactly once. Furthermore, by juggling $juggle(r)$ with bell balls such that throws 1 to b in $juggle(r)$ are executed with balls 1 to b, respectively, we ring r over and over again. For example, if r is the 3-bell extent Plain Bob Singles in our example, then

$$juggle(\text{Plain Bob Singles}) = (423\ 342)^3.$$

As in this example, we will usually highlight the changes of $juggle(r)$ by separating adjacent ones by a space.

On a practical note, we remark that 423342 is not a difficult sequence to juggle if, as usual, we treat the 2-throws as holds. However, since the bell ball sounds are triggered by the balls being caught by the hands, we really have to execute these 2-throws, which is not as easy as it sounds. To be able to do this, you either end up throwing the 4-throws really high or you settle for bouncing the 2-throws off one hand (these throws are only performed by one hand).

We summarize the most important properties of juggling sequences that arise from ringing sequences as follows:

Juggling Sequences From Ringing Sequences

Let r be a ringing sequence on $b \geq 2$ bells consisting of at least two changes. Then, the following hold:

(R1) The changes of $juggle(r)$ and $juggle(r)$ itself are ground-state juggling sequences.

(R2) The throws in $juggle(r)$ are to height $b-1$, b, and $b+1$ only.

The second property is an immediate consequence of the fact that from one change to the next a bell moves by at most one position in its order of ringing.

Remember the problem that a ring of bells is a magnificent musical instrument that cannot be used to play tunes if we are not prepared to compromise in terms of the quality of the sound produced? The most popular solution to this problem that people have come up with is change ringing.

For bell balls, we have a corresponding problem and solution. On the one hand, it is possible to translate any tune into a juggling sequence in pretty much the same way as we translated ringing sequences (with some obvious modifications). This means that, in theory, we can juggle tunes. However, most tunes translate into juggling sequences that feature throws to many different heights. To juggle such a sequence is very hard in practice. The solution to this problem is to juggle the changes because then we only have to worry about three different "doable" throw heights: $b - 1$, b, and $b + 1$.

Note that any b in $juggle(r)$ corresponds to a bell that does not move its position from one of the changes to the next. Composers of ringing sequences that are actually performed try to minimize the number of times this happens. This means that, for large b, given such a ringing sequence r on b bells, the number of bs in $juggle(r)$ will be small compared to the number of $(b-1)$s and $(b+1)$s. Because of this and the fact that transitions from one change of r to the next are effected by swapping positions of adjacent bells, you will find that $juggle(r)$ is made up of (large if b is large) blocks of the form $((b + 1)(b - 1))^k$ separated by small blocks of the form b^j. For example, $juggle(r)$ of an extent of Plain Bob Major (an extent on eight bells) starts as follows:

$$97889797\ 89789797\ 97978897\ 97889797\ 97978897\ldots.$$

This means that the best way to prepare for juggling ringing sequences of b bells is first to learn to juggle the sequences $(b+1)(b-1)$ (a half-shower in the case of b even), $(b + 1)(b - 1)b$, and $(b + 1)(b - 1)b^2$. For some of these sequences, this has to be done in both directions since these sequences are not symmetrical with respect to the left- and right-hand actions. Also very important is to practice the transitions between these sequences and between these sequences and the basic b-ball sequence. Note that these sequences are among the first sequences anybody interested in performing b-ball juggling sequences will be learning anyway.

As with bells, juggling the changes becomes interesting and challenging only once you try to do this with more than three balls. In practice, I think most good jugglers should be able to juggle extents on four balls; Table 6.3 lists the most popular ones. Every such juggling sequence has period $4 \cdot 24 = 96$. In terms of memorization and execution, the extents of Erin and Reverse Erin listed in the table are probably the easiest 4-bell extents; see the next sections for more information about these particular extents. Some shorter ringing sequences and maybe even full extents on

name	juggling sequence
Plain Bob	$(5353\ 4534\ 5353\ 4534\ 5353\ 4534\ 5353\ 4453)^3$
Reverse Bob	$(5353\ 4534\ 5353\ 5344\ 5353\ 4534\ 5353\ 4534)^3$
Double Bob	$(5353\ 4534\ 5353\ 5344\ 5353\ 4534\ 5353\ 4453)^3$
Canterbury	$(5353\ 4534\ 4453\ 5344\ 4453\ 4534\ 5353\ 4534)^3$
Reverse Canterbury	$(5344\ 4534\ 5353\ 4534\ 5353\ 4534\ 5344\ 4453)^3$
Double Canterbury	$(5344\ 4534\ 4453\ 5344\ 4453\ 4534\ 5344\ 4453)^3$
Single Court	$(5344\ 4534\ 5353\ 4534\ 5353\ 4534\ 5344\ 4534)^3$
Reverse Court	$(5353\ 4534\ 4453\ 4534\ 4453\ 4534\ 5353\ 4534)^3$
Double Court	$(5344\ 4534\ 4453\ 4534\ 4453\ 4534\ 5344\ 4534)^3$
St Nicholas	$(5344\ 4534\ 5353\ 5344\ 5353\ 4534\ 5344\ 4453)^3$
Reverse St Nicholas	$(5353\ 4534\ 4453\ 5344\ 4453\ 4534\ 5353\ 4453)^3$
Erin	$(5344\ 4534\ 5344\ 4534\ 5344\ 5353)^4$
Reverse Erin	$(4453\ 4534\ 4453\ 4534\ 4453\ 5353)^4$
Stanton	$(5344\ 4534\ 4453\ 4534\ 5344\ 5353)^4$
Reverse Stanton	$(4453\ 4534\ 5344\ 4534\ 4453\ 5353)^4$

TABLE 6.3. The juggling sequences corresponding to the eleven method-based 4-bell extents and some principle-based 4-bell extents.

five balls consisting of $5 \cdot 120 = 600$ throws may be within the reach of dedicated jugglers. At what speed would we be juggling the changes? Well, as for me, with four balls I can juggle in the range of 40–60 changes per minute, which sounds good.

Of course, if you have a set of bell balls, it is also fun to just juggle your favorite patterns. One of the best things about bells is that, no matter how you ring them, they will almost always sound good.

6.2.2 Turning Extents into Site Swaps

There are $b!$ different changes in an extent on b bells. There are the same number of b-ball (simple) ground-state juggling sequences of period b; see Subsection 2.8.2. In [94], Martin Probert pointed out the following connection. Remember that to get from one change of a ringing sequence to the next, we swap the positions of the bells in one or more pairs of adjacent bells starting with rounds. Now, Probert's idea is to replace rounds by the basic b-ball juggling sequence of period b and, in general, the changes in the ringing sequence by the b-ball ground-state juggling sequences of period b. Instead of a number of "swaps" to move from one change to the next, we now use site swaps in the respective positions to move from juggling sequence to juggling sequence. As an example, consider how everybody's favorite 4-bell extent of Plain Bob Minimus translates into a sequence of 4-ball ground-state juggling sequences; see Table 6.4.

Here is why this always works no matter what extent we translate. Let s be any b-ball ground-state juggling sequence of period b. Then, we may

1234	1342	1423		4444	4552	4633
2143	3124	4132		5353	6334	7342
2413	3214	4312		5623	6424	7522
4231	2341	3421		7441	5551	6631
4321	2431	3241		7531	5641	6451
3412	4213	2314		6622	7423	5524
3142	4123	2134		6352	7333	5344
1324	1432	1243		4534	4642	4453

TABLE 6.4. The extent of Plain Bob Minimus and the corresponding sequence of 4-ball simple juggling sequences containing all 4-ball simple ground-state juggling sequences of period 4 exactly once.

assume that its ith throw gets executed on beat i. Now, the important thing is that when we juggle this sequence, any of the balls that is thrown on any of the beats from 1 to b lands after the bth beat. Consequently, we can apply site swaps to any of the pairs of elements in the sequence and, after such a site swap, we are dealing with a new ground-state sequence. So, basically, we can do whatever we like in terms of site swaps. This means that we can actually reinterpret an extent in this way in the first place, that we will end up with the right number of b-ball ground-state juggling sequences, and that because in doing so we are cycling through all possible permutations of throwing sequences, all generated ground-state sequences will be distinct.

Of course, it should be clear that for this "application" of ringing sequences we do not need any bell balls. Also, when juggling balls, we are not restricted to swapping adjacent balls. This immediately suggests a couple of things to try. For example, in [94], Probert gives the following way to play through all the 4-ball ground-state juggling sequences of period 4 with only two throws changing as we pass from sequence to sequence:

4444	5641	6451	7342	6622	5524
4453	5551	6631	7333	6424	5623
4552	5353	7531	7423	6334	4633
4642	6352	7441	7522	5344	4534

TABLE 6.5. Playing through all 4-ball simple ground-state juggling sequences of period 4 with two throws changing at every step.

6.3 Mathematical Notation and Basic Operations

In the following, we give an introduction to the mathematics of bell ringing. We start by fixing the mathematical notation that we will be using for the rest of this chapter and deriving some basic operations that can be used to transform ringing sequences into new ringing sequences.

6.3.1 Notation

Changes in a b-bell ringing sequence "are" permutations of the ordered set $[b]$ consisting of the integers from 1 to b. As such, they can be considered as elements of the symmetric group S_b. The transitions from one change to the next can also be considered as elements of this symmetric group. Because of Property B3 (see page 142), these elements are special involutions; that is, special elements of order 2. Although it is not necessary to distinguish between permutations that are changes and permutations that are transitions between changes, we will make this distinction for clarity of exposition's sake. To also visually distinguish between the two, we will note changes, as we have done so far, in the notation that bell ringers use and transitions in cycle notation preferred by group theorists. For example,

$$123456$$

stands for the change "rounds" on six bells and

$$(23)(45)$$

denotes a transition that specifies that the positions of the bells in the second and third positions swap places as well as those in the fourth and fifth positions. Applying this transition to rounds on six bells, as above, yields the change

$$132546.$$

This means that any b-bell ringing sequence of length $n+1$ corresponds to a (b-bell) *transition sequence* of length n that records the transitions from change to change starting with rounds. We will note such a sequence, very much like a juggling sequence, as a word of length n, every letter of which stands for a particular kind of transition. For example, there are exactly two transitions that are used to construct all 3-bell ringing sequences. We denote these by

$$A = (12),$$
$$B = (23).$$

Then, the transition sequence that corresponds to the 3-bell extent in Table 6.1 is

$$ABABAB = (AB)^3.$$

Sometimes, we will also consider a transition sequence as a product of elements of a certain symmetric group. *To evaluate such a product, we proceed, as is customary, from right to left.* Here is another important remark.

Let c be a change, T be a transition, and d be the change that we arrive at by applying T to c. If we consider all three elements as elements of the respective symmetric group, then d in fact turns out to be the product cT (again evaluated from right to left).

6.3.2 Ringing Sequences from Ringing Sequences

In this subsection, we introduce some of the most important and useful procedures for turning a b-bell ringing sequence r into a new ringing sequence. It is usually easiest to describe these procedures in terms of the *ringing array* of r. Remember that if $l+1$ is the length of r, then the ringing array of r is the $b \times (l+1)$ array whose ith row is the ith change of the ringing sequence. Finally, let

$$t = A_1 A_2 \cdots A_l$$

be the transition sequence that corresponds to r.

Inverse, Reverse, and Cyclic Shift

The *inverse* of r is the b-bell ringing sequence whose first change is the last change of r, whose second change is the second to last change of r, and so forth. In terms of ringing arrays, we basically reflect at a horizontal axis to move from r to its inverse.

If we replace every element of the ringing array of r by the number $(b+1-$this element$)$ and then reflect the resulting array at a vertical axis, we arrive at a new ringing array, which we call the *reverse* of r. Clearly, the reverse of the reverse of r and the inverse of the inverse of r coincide with r.

The *cyclic shift* of r is the b-bell ringing sequence that corresponds to the transition sequence

$$A_l A_1 A_2 \cdots A_{l-1}.$$

Vertical Shifts and Elevation

There are also some ways to turn the b-bell ringing sequence r into $(b+k)$-bell ringing sequences, where k is a positive integer.

First, note that the transition sequence of r is also the transition sequence of a $(b+k)$-bell ringing sequence, which we call the *right vertical k-shift*.

Now, add k to every entry of the ringing array of r and prefix every row of the resulting array by $1234\cdots k$. Then, this new array is also the ringing array of a $(b+k)$-bell ringing sequence, which we call the *left vertical k-shift* of r.

In the special case where l is even, White [146] introduced a neat procedure for turning the b-bell ringing sequence r into a $(b+1)$-bell ringing sequence that is an extent if and only if r is. We call this new sequence the *elevation* of r. Here is how it is constructed. Remember that our ringing

2-bell extent r	auxiliary array	elevation of r
12	23	123
21	23	213
	23	231
	32	321
	32	312
	32	132

TABLE 6.6. Turning the 2-bell extent into a 3-bell extent.

3-bell extent r	auxiliary array			elevation of r		
123	234	342	423	1234	1342	1423
213	234	342	423	2134	3142	4123
231	234	342	423	2314	3412	4213
321	234	342	423	2341	3421	4231
312	324	432	243	3241	4321	2431
132	324	432	243	3214	4312	2413
	324	432	243	3124	4132	2143
	324	432	243	1324	1432	1243

TABLE 6.7. Turning a 3-bell extent into a 4-bell extent.

sequence r has length $l + 1$. We start by constructing a $b \times ((b + 1)l + 1)$ array whose first $b + 1$ rows coincide with the first change of r, the second $b + 1$ rows with the second change of r, and so on. We now add 1 to every entry of the resulting array to arrive at what we call an *auxiliary array*. Following this, we squeeze an additional 1 into every row of the auxiliary array such that in rows 1,2,3, ... of the new array this 1 occupies positions

$$1, 2, \dots, b, b + 1, b + 1, b, \dots, 1, 1, 2, \dots, b, b + 1, b + 1, b, \dots, 1, \dots,$$

respectively. This new array is the ringing array of the $(b + 1)$-ringing sequence we are after. By iterating this construction, we can construct extents on any number of bells starting with the extent on one bell. See Tables 6.6 and 6.7 for details of the constructions of the resulting 3-bell and 4-bell extents. As usual, we have omitted the last rows of all the arrays in these tables. We remark that the elevation of the 3-bell extent Plain Bob Singles, as calculated in Table 6.7, is the 4-bell extent Double Canterbury Minimus. The elevation of the inverse of Plain Bob Singles is the 4-bell extent Double Court Minimus. Both 4-bell extents are very nicely behaved from both bell ringing and mathematical points of view, as we shall see in the following sections. For a detailed analysis of the different b-bell extents that arise from the 1-bell extent by successive "elevation," see [146].

We summarize the results of this subsection as follows:

Operations on Ringing Sequences

Let r be a b-bell ringing sequence or extent of length $l + 1$. Then, the following hold:

(O1) The inverse, reverse, and cyclic shift of r are b-bell ringing sequences or extents, respectively.

(O2) A vertical 1-shift or elevation of r (if l is even) is a $(b + 1)$-bell ringing sequence. The elevation of r is an extent if and only if r is an extent.

(O3) For any positive integer b, there exists a b-bell extent.

6.4   Principles and Methods

Most extents that are actually rung by change ringers are based on special sequences of transitions called *methods* and *principles*. In the following, we define these two terms and describe the overall structure of the extents built from these basic building blocks. The following definitions may seem rather technical and unintuitive at first. However, a couple of examples should clarify things.

Let t be a b-bell transition sequence, where b is at least 3, and let m be the permutation you arrive at by multiplying all elements in t in the order that they appear in this sequence (from right to left). This is an element of the symmetric group S_b. Let o be the order of m, and assume that t^o is the transition sequence of a b-bell ringing sequence p. Then, p is called the *plain course* of t. The plain course (minus rounds at the end) has a natural subdivision into o blocks, called *leads*, of the same length as t itself. The *lead ringing array* of this plain course has as entries the changes of the plain course (minus the last change). Its o columns are the o blocks of the plain course in their natural order.

If the permutation m fixes some bells, then change ringers say these bells are not *working*, while the bells that are not fixed are said to be working.

6.4.1   Principles

Our original transition sequence t is called a *principle* if the permutation m is a cycle of order $o = b$ or, as bell ringers would express it, all bells are working. For example, let's consider the 4-bell principle Erin Minimus

$$(DB)^2 DA,$$

where $A = (12)(34)$, $B = (23)$, and $D = (12)$. In this example, $m = (1342)$ and we see that the plain course of Erin Minimus is a 4-bell extent.

1234	3142	4321	2413
2134	1342	3421	4213
2314	1432	3241	4123
3214	4132	2341	1423
3124	4312	2431	1243
1324	3412	4231	2143

TABLE 6.8. The lead ringing array of the plain course of the 4-bell principle Erin Minimus. The four columns of this array are the four leads of the plain course. The plain course itself is a 4-bell extent. The path of bell 1 in lead 1 is mirrored exactly by the paths of bells 3, 4, and 2 in leads 2, 3, and 4, respectively.

Change ringers express the fact that the permutation m is a cycle of order b by saying that *in a principle all bells do the same work*, meaning that the path that bell i follows through lead j of the plain course is precisely followed by bell $m(i)$ in lead $j + 1$ for $j + 1 \leq o$ and lead 1 for $j + 1 > o$. For example, in Table 6.8, the path of bell 1 in lead 1 is followed precisely by bell 3 in lead 2, by bell 4 in lead 3, and by bell 2 in lead 4. The path of bell 1 in lead 2 is followed by bell 2 in lead 1, by bell 3 in lead 3, and by bell 4 in lead 4, and so forth.

6.4.2   Methods

Assume that the treble (bell 1) and possibly some other bells do not work; that is, they are fixed by m. Our original transition sequence t is called a *method* if the permutation m is a cycle whose order is the number of working bells and the bells that do not work *hunt*. Roughly speaking, for a bell to hunt means that in the plain course it weaves back and forth across the ringing array in one of a number of different more or less intricate ways bearing such names as Plain Hunt, Treble Dodging, Treble Place, Alliance, and Hybrid.

For our purposes, it will be sufficient to assume that in a method all bells that do not work *plain hunt*. In particular, for the treble to plain hunt means that in successive changes it occupies positions

$$1, 2, \ldots, b, b, b - 1, \ldots, 1, 1, 2, \ldots, b, b - 1, \ldots, 1, \ldots .$$

The other nonworking/plain hunting bells in the method follow the same path commencing at an appropriate spot. As an example, let us consider the 5-bell method Grandsire Doubles,

$$FBA(FB)^2 FABFB,$$

where $A = (12)(34)$, $B = (23)(45)$, and $F = (12)(45)$. Table 6.9 shows the lead ringing array of the plain course, and we see that there are exactly two plain hunting bells, namely bells 1 and 2. Note that this plain course is not

12345	12534	12453
21354	21543	21435
23145	25134	24153
32415	52314	42513
34251	53241	45231
43521	35421	54321
45312	34512	53412
54132	43152	35142
51423	41325	31524
15243	14235	13254

TABLE 6.9. The lead ringing array of the plain course of the 5-bell method Grandsire Doubles. The three columns of this array are the leads of the method. Both bells 1 and 2 hunt. The plain course is not an extent.

an extent. As in this example, the length of a b-bell method is always $2b$. Furthermore, as in principles, all working bells do the same work.

	name	transition sequence
methods	Plain Bob	$(AB)^3 AC$
	Reverse Bob	$ABAD(AB)^2$
	Double Bob	$ABADABAC$
	Canterbury	$ABCDCBAB$
	Reverse Canterbury	$DB(AB)^2 DC$
	Double Canterbury	$DBCDCBDC$
	Single Court	$DB(AB)^2 DB$
	Reverse Court	$AB(CB)^2 AB$
	Double Court	$DB(CB)^2 DB$
	St. Nicholas	$DBADABDC$
	Reverse St. Nicholas	$ABCDCBAC$
principles	Erin	$(DB)^2 DA$
	Reverse Erin	$(CB)^2 CA$
	Stanton	$DBCBDA$
	Reverse Stanton	$CBDBCA$

TABLE 6.10. The transition sequences corresponding to all eleven 4-bell methods and four 4-bell principles generated by three transitions each. The transitions are $A = (12)(34)$, $B = (23)$, $C = (34)$, and $D = (12)$.

A hunting treble in a ringing sequence helps ringers keep track of where exactly in the ringing of a sequence they are at any given moment.

There is exactly one 3-bell method (see its lead ringing array in Table 6.1) and eleven 4-bell methods; see Table 6.10. In the 3-bell method, all bells hunt, whereas in the 4-bell methods only the treble hunts. This means that the order o of the permutation m associated with 3- and 4-bell methods is 1 and 3, respectively. Hence, the lengths $2ob + 1$ of the corresponding plain courses are 7 and 25, respectively; that is, the plain courses are extents.

As in the case of 3- and 4-bell methods, the plain courses of some 4-bell principles are extents; Table 6.10 lists the four essentially different 4-bell principles made up of only three different transitions each of whose plain courses are extents (three is the minimum number of transitions required to construct such a principle).

6.4.3 Extents Based on Principles or Methods

A transition sequence e is the transition sequence of a *b-bell extent of a method or principle* t if it can be written in the form

$$e = t_1 t_2 \cdots t_k,$$

where the t_is are transition sequences of the same length as t that differ from t in (usually) at most one element each. Change ringers refer to such distinguished elements as *calls*. As we have seen above, it is possible that the extent coincides with the plain course of t. However, in the case of t being a method, it is easy to conclude from what we said before that this is only possible if the number of bells is 3 or 4.

In the actual ringing of an extent of a method or principle t, every one of the ringers memorizes t and only one of the ringers, called the *conductor*, also needs to memorize the positions in the extent where a call needs to be executed instead of the usual transition at the given moment. This means that to ring the extent starting with rounds, the ringers ring t over and over until they either arrive back at rounds, in which case the plain course of t coincides with the full extent, or the conductor "makes a call." After making the corresponding adjustment, the ringers return to ringing t until they arrive back at rounds or the conductor makes another call, and so on.

	Plain Bob Doubles	Stedman Doubles
t	$(AB)^4 AC$	$FBA(FB)^2 FABFB$
e	$(((AB)^4 AC)^3 (AB)^4 AD)^3$	$((FBA(FB)^2 FABFB)^4$
		$FBA(FB)^2 FABFD)^2$
m	(2354)	(15234)
calls	D (three times)	D (twice)

TABLE 6.11. The structure of two popular 5-bell extents. The transitions used are $A = (12)(34)$, $B = (23)(45)$, $C = (34)$, $D = (23)$, and $F = (12)(45)$.

Table 6.11 lists for extents of the popular 5-bell method Plain Bob Doubles and the 5-bell principle Stedman Doubles the transition sequences e and t as well as the cycle m (which permutes the working bells) and the different calls used. In both extents, calls are the last elements of leads. Table 6.12 exhibits the structure of an extent based on the method Grandsire

Doubles. In this extent, two different calls are used. Note that the call F is also used in the method itself and is made as the second to last element of a lead.

	Grandsire Doubles
t	$FB(AB)^4$
e	$((FB(AB)^4$
	$\quad FB(AB)^3 FB)^2$ (the last A in the method gets replaced by F)
	$FB(AB)^4$
	$FB(AB)^3 FG)^2$ (the last B in the method gets replaced by G)
m	(354)
calls	F (four times) and G (twice)

TABLE 6.12. The structure of a popular 5-bell extent of the method Grandsire Doubles. The transitions are the same as in Table 6.11, with the addition of the transition $G = (45)$.

As in our examples, to further facilitate memorization of ringing sequences of methods and principles, calls are usually made near the ends of leads, and there are only very few different calls that are used to construct any given extent. Similarly, the methods and principles that are actually rung are usually based on very few different transitions. Composers of extents will also try to keep the total number of calls in an extent as small as possible. Note that if $l + 1$ is the length of the plain course of t, then at least $b!/l - 1$ calls are needed to compose an extent of t.

In ringing sequences of methods, calls are usually chosen such that they do not affect the path of the treble. This means that in ringing sequences of methods, the treble usually plain hunts throughout.

Many methods and principles are *palindromic*, meaning that if

$$A_1 A_2 \cdots A_{k+1}$$

is such a method or principle, then $A_1 = A_k$ (not $A_1 = A_{k+1}$), $A_2 = A_{k-1}$, and so on. Verify that most examples of methods and principles in this chapter have this property.

We should also mention that the *Decisions of the Central Council of Church Bell Ringers* [27] are a complete system of rules that change ringers worldwide abide by both in the actual ringing and in the composition of ringing sequences. Apart from the rules and guidelines mentioned above, the Decisions also require that "no bell shall make more than four consecutive blows in the same position, this requirement not applying to Minimus methods." In fact, composers will usually try to keep the number of times that a bell stays in a position for more than two successive changes to a minimum.

6.5   Graphical Representations of Extents

In the following, we give interpretations of extents in terms of graphs and use these to highlight the beauty of some of the most popular extents on three to five bells.

6.5.1   Cayley Graphs

The so-called *b-bell Cayley graph* has as its vertices the different *b*-bell changes, and two of its vertices are connected by an edge labeled by one of the (*b*-bell) transitions if and only if this transition is a transition between the two vertices. This means that this graph has *b*! vertices. The number of different labels of its edges—that is, the number of transitions in the symmetric group S_b—is $F(b) - 1$, where $F(b)$ is the *b*th *Fibonacci number*; see [144], Theorem 2.1, for a proof of this fact. Remember that $F(b)$ is defined recursively for any nonnegative integer *b* by setting

$$F(0) = F(1) = 1, \quad F(b+1) = F(b) + F(b-1).$$

Hence, $F(2) - 1 = 1$, $F(3) - 1 = 2$, $F(4) - 1 = 4$, $F(5) - 1 = 7$, and so on. There is exactly one edge of every kind ending at every vertex.

To avoid a cluttered appearance in graphical representations of the *b*-bell Cayley graph and its subgraphs, we only label the edges but not the vertices of this graph. Figure 6.3 shows the 3-bell Cayley graph with unlabeled vertices.

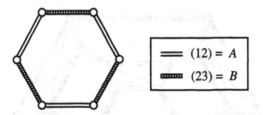

FIGURE 6.3. The 3-bell Cayley graph.

Note that if we do not label the vertices of this graph, none of these vertices is distinguished in any way among the rest; that is, the automorphism group of the graph acts transitively on the vertices. In fact, we do not lose any information about the graph by not labeling the vertices. To reconstruct a valid labeling of the vertices by changes, start by labeling an arbitrary vertex *s* (for "start") with an arbitrary change. Given any other vertex *e* (for "end"), choose a path in the Cayley graph that connects *s* with *e*. As you travel from *s* to *e*, successively apply the transformations that correspond to the edges that make up the path to turn the label for *s* into the label for *e*.

Now, it is easy to see that the following hold; see also [144], Theorem 2.2:

Extents are Oriented Hamiltonian Cycles

The b-bell ringing sequences with $b \geq 3$ and length at least 4 correspond precisely to the oriented *cycles* in the b-bell Cayley graph through a fixed vertex (that is, the oriented circular paths through this vertex that contain every vertex at most once). A b-bell ringing sequence is an extent if and only if its corresponding cycle is *Hamiltonian* (that is, it contains every vertex exactly once).

Every cycle in the Cayley graph corresponds to two oriented cycles. Furthermore, the two ringing sequences corresponding to these two oriented cycles are inverses of each other. For example, if we label the upper-left vertex in the 3-bell Cayley graph in Figure 6.3 by rounds, we see that there is exactly one Hamiltonian cycle in the graph that starts at this vertex. The two corresponding oriented cycles translate into the transition sequences $(AB)^3$ and $(BA)^3$ of the 3-bell extent in Table 6.1 and its inverse.

6.5.2 Four Bells

Figure 6.4 shows a rendering of the 4-bell Cayley graph that features an order 3 rotational symmetry.

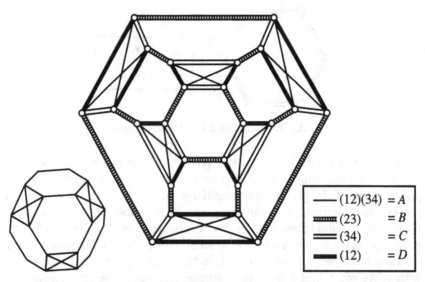

—— (12)(34)	= A
▫▫▫▫ (23)	= B
══ (34)	= C
▬▬ (12)	= D

FIGURE 6.4. The 4-bell Cayley graph is a truncated octahedron with "crosses across its square faces."

Figure 6.5 shows Hamiltonian cycles corresponding to the plain courses of the four 4-bell principles in Table 6.10.

Figure 6.6 shows Hamiltonian cycles in this graph corresponding to the plain courses of the eleven 4-bell methods listed in Table 6.10. As a starting vertex, we have chosen the upper-left vertex of the inner hexagon of the Cayley graph. Remember that all these plain courses are 4-bell extents that have divisions into three leads of eight changes each. These order 3 symmetries of the extents translate into order 3 rotational symmetries of the corresponding cycles. The fact that all eleven methods are palindromic translates into order 2 (mirror) symmetries of the Hamiltonian cycles. Note that some of the cycles actually look like the union of three bells. Of course, this is coincidence.

The most symmetric spatial rendering of the 4-bell Cayley graph has as its vertices those of a truncated octahedron and as edges the suitably labeled edges of this Archimedean solid plus the diagonals of its square faces; see again Figure 6.4 (lower-left corner).

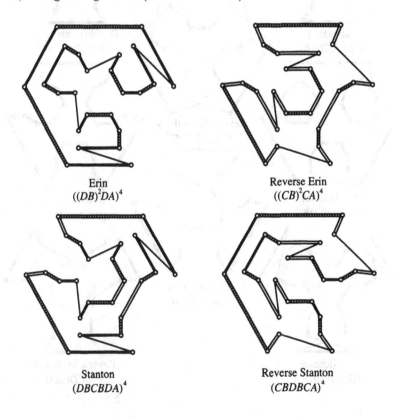

Erin
$((DB)^2DA)^4$

Reverse Erin
$((CB)^2CA)^4$

Stanton
$(DBCBDA)^4$

Reverse Stanton
$(CBDBCA)^4$

FIGURE 6.5. Hamiltonian cycles in the 4-bell Cayley graph that correspond to the plain courses of four 4-bell principles. The common starting vertex is the upper-left vertex of the inner hexagon in Figure 6.4.

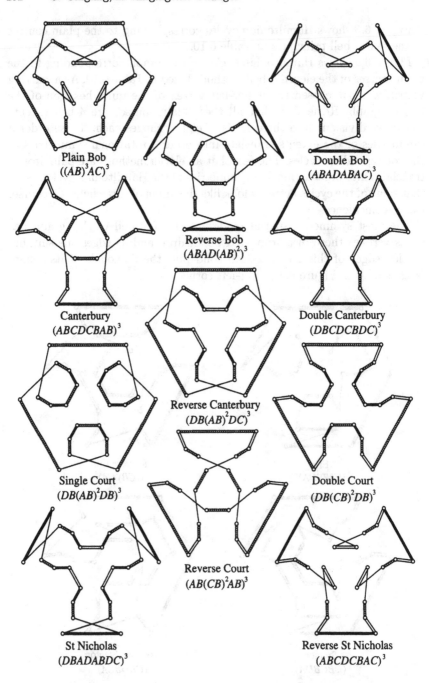

Plain Bob
$((AB)^3AC)^3$

Double Bob
$(ABADABAC)^3$

Reverse Bob
$(ABAD(AB)^2)^3$

Canterbury
$(ABCDCBAB)^3$

Double Canterbury
$(DBCDCBDC)^3$

Reverse Canterbury
$(DB(AB)^2DC)^3$

Single Court
$(DB(AB)^2DB)^3$

Double Court
$(DB(CB)^2DB)^3$

Reverse Court
$(AB(CB)^2AB)^3$

St Nicholas
$(DBADABDC)^3$

Reverse St Nicholas
$(ABCDCBAC)^3$

FIGURE 6.6. Hamiltonian cycles in the 4-bell Cayley graph that correspond to the plain courses of the eleven methods on four bells. The common starting vertex is the upper-left vertex of the inner hexagon in Figure 6.4.

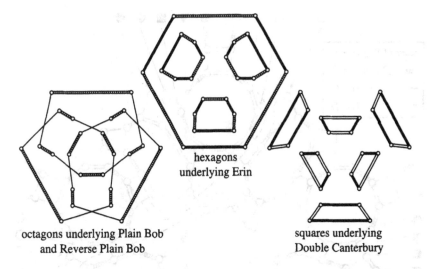

hexagons
underlying Erin

octagons underlying Plain Bob
and Reverse Plain Bob

squares underlying
Double Canterbury

FIGURE 6.7. Partitions of the 4-bell Cayley graph underlying various extents.

Although the overall structure of the extents is very apparent in the 2-dimensional diagrams, they do not give maximum insight into the overall symmetries of the underlying extents. Embedded in the 3-dimensional model of the Cayley graph, the Hamiltonian cycles corresponding to the eleven methods and four principles display most of the mathematical beauty and symmetries of the corresponding extents; see the Appendix for stereograms of the 3-dimensional models of these Hamiltonian cycles and instructions on how to view them.

Consider, for example, the extents of Plain Bob, Double Canterbury, and Erin. These extents are based on "partitions" of the Cayley graph into disjoint octagons, squares, and hexagons, respectively, whose edges are labeled with two different transitions each. Whereas in the 3-dimensional model these partitions "jump out" of the picture, in the 2-dimensional case some highlighting is in order; see Figure 6.7. To construct the Hamiltonian cycles from the partition, we remove one edge from each of the polygons and splice the resulting open paths together using edges corresponding to one particular kind of transition. We note that in the 3-dimensional model, the octagons underlying Plain Bob (and Reverse Plain Bob) are the three "great circles" of the truncated octahedron and the hexagons and squares border faces of this Archimedean solid.

6.5.3   Five Bells

A number of very attractive Hamiltonian cycle representations of 5-bell extents can be found in [144]. Here, we only present one such representation of an extent of the method Plain Bob Doubles that reflects very nicely the overall structure of such an extent.

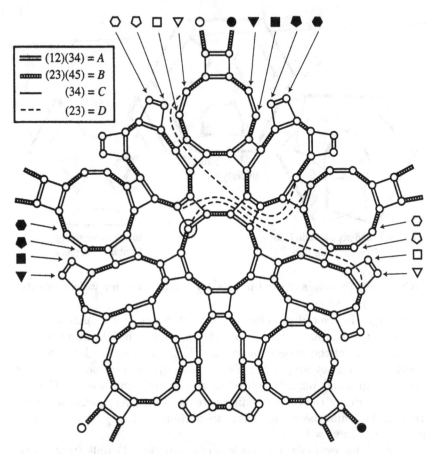

FIGURE 6.8. A subgraph of the 5-bell Cayley graph used to picture the extent of Plain Bob Doubles $(((AB)^4AC)^3(AB)^4AD)^3$.

The extent is given by the transition sequence

$$(((AB)^4AC)^3(AB)^4AD)^3,$$

where A, B, C, and D are as in Figure 6.8. Note that Plain Bob Doubles is the transition sequence

$$(AB)^4AC,$$

and the transition sequence corresponding to its plain course is

$$((AB)^4AC)^4.$$

There are seven different transitions in the symmetric group S_5. This means that in the full 5-bell Cayley graph every one of its 120 vertices is the end point of seven different edges. This makes it impractical to actually draw the whole graph. However, note that most of the transitions in the transition sequence above are As, Bs, and Cs; there are only three Ds. This

means that to get a good idea of how this extent works, we can restrict
ourselves to drawing the subgraph of the full 5-bell Cayley graph that
consists of all its vertices and all the edges that are labeled with an A,
a B, or a C. Figure 6.8 shows a rendering of this subgraph that features
an order 5 symmetry. Certain points on the boundary of this graph are
identified in pairs as indicated. Note that we only include one essential
part of the pairs that get matched up. The remaining pairs can be found
by exploiting the order 5 rotational symmetry of the diagram. We also
include a starting point for our cycle (the circled vertex) and three edges
labeled D. Starting at the distinguished vertex, it is now possible to plot
the complete Hamiltonian cycle that corresponds to the transition sequence
under discussion. Figure 6.9 shows this cycle. The pairs of numbers indicate
how the disconnected components of this cycle fit together.

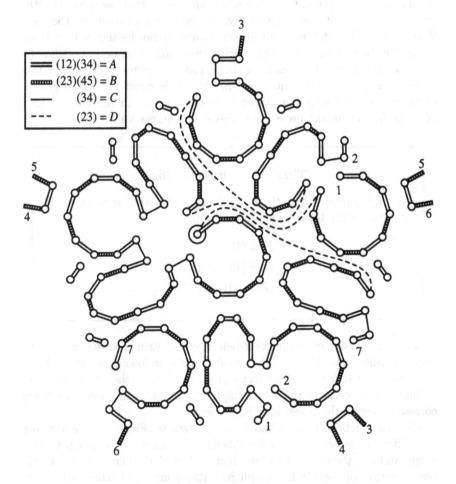

FIGURE 6.9. A Hamiltonian cycle in the 5-bell Cayley graph.

In Figure 6.9, the overall structure of the transition sequence is apparent. At the lowest level, we "see" the method (minus the C) in terms of the twelve 10-gons that get strung together to make up the whole cycle. On a higher level, it becomes apparent that the cycle is really just three plain courses of the method "spliced together" by three Ds. Follow the first plain course from the starting vertex, and you see that you really would get back to this vertex prematurely if no call was made at the appropriate time replacing the usual C by a D.

6.5.4 Many Bells

In the case of two and three bells, we have seen that we need exactly one and two transitions, respectively, to generate an extent. In the case of four bells, three different transitions suffice. It can be shown that if we are dealing with more than three bells, two transitions do not suffice to compose an extent. The reason for this is that the group generated by two involutions in the symmetric group S_b, $b \geq 4$, is never the whole group. However, we have seen that three transitions suffice in the case of four bells. In [97], Rapaport proves that this is true in general.

Three Transitions Suffice

For any integer $b \geq 4$, there is a b-bell extent that is constructed using only the three transitions

$$E = (12),$$
$$F = (12)(34)(56)\cdots,$$
$$G = (23)(45)(67)\cdots.$$

We have already seen that the 4-bell extent of Erin is constructed using only the three transitions in this result. In the following, we may therefore assume that $b > 4$. Rapaport's proof consists in the construction of a Hamiltonian cycle in the b-bell Cayley graph that only contains edges corresponding to the transitions E, F, and G.

The construction itself is based on two results. First, if X and Y are two different transitions used for labeling the edges of the b-bell Cayley graph and o is the order of the product of X and Y, then there is exactly one partition of the Cayley graph into $(2o)$-gons all of whose edges are labeled X and Y; see, for example, the partitions of the 4-bell Cayley graph in Figure 6.7. Here is the second building block for Rapaport's result:

A connected regular graph of degree 3 has a Hamiltonian cycle if there is a set P of polygons and a set Q of squares, each set partitioning the vertex set of the graph and such that no member of P contains every vertex of a member of Q.

In our case, the graph consists of all the vertices and the edges labeled E, F, and G of the b-bell Cayley graph. The set of polygons is composed of the 12-gons associated with the transitions E and G (or we could use the $(2b)$-gons associated with the transitions F and G). The set Q consists of the squares associated with the transitions E and F. To construct a Hamiltonian cycle, Rapaport then uses F edges to combine 12-gons from P (or E edges to combine $(2b)$-gons from P).

In [144], Section 3, White investigates the properties of Rapaport's set of three transitions and similar sets of three transitions that can be used to generate b-bell extents.

6.5.5 Names

Here are some remarks about what the names of the methods and principles that we encountered thus far have to do with their structure and the relationships among them.

In general, the reverse of the plain course of a b-bell principle, as we defined it in Subsection 6.3.2, is always the plain course of another principle. Let

$$A_1 A_2 \cdots$$

be the transition sequence of the first principle where

$$A_i = (a_{1,i}, a_{2,i})(a_{3,i}, a_{4,i}) \cdots .$$

Then, the transition sequence of the second principle is

$$\overline{A}_1 \overline{A}_2 \cdots ,$$

where

$$\overline{A}_i = (b+1-a_{1,i}, b+1-a_{2,i})(b+1-a_{3,i}, b+1-a_{4,i}) \cdots .$$

For example, in the case of the transition sequence $(DB)^2 DA$ of Erin, we find

$$\overline{(DB)^2 DA} = (CB)^2 CA,$$

which is the transition sequence for Reverse Erin.

In general, the reverse of the plain course of a b-bell method is not the plain course of another method. However, the bth cyclic shifts of this reverse will always be the plain course of a method. Furthermore, if t is the transition sequence of the first method, then b cyclic shifts of $\overline{t}$ give the transition sequence of the second method, which we then call Reverse First Method. For example, in the case of the transition sequence $(AB)^3 AC$ of Plain Bob, we find that

$$\overline{(AB)^3 AC} = (AB)^3 AD,$$

four cyclic shifts of which give

$$ABAD(AB)^2,$$

which is Reverse Plain Bob. We find similarly for Canterbury and Reverse Canterbury and Single Court and Reverse (Single) Court.

It should be quite obvious by looking at the diagrams how Plain Bob and Reverse Plain Bob have been combined into Double (Plain) Bob. Here the Double stands for the fact that Double (Plain) Bob = Reverse Double (Plain) Bob; just check that four cyclic shifts of

$$\overline{ABADABAC} = ABACABAD$$

equals

$$ABADABAC$$

and similarly for Double Canterbury and Double Court.

Finally, again by inspecting the diagrams in Figure 6.6 and the corresponding stereograms in the Appendix, it should be obvious that St. Nicholas is a combination of Reverse Plain Bob and Reverse Canterbury, whereas Reverse St. Nicholas is a combination of Plain Bob and Canterbury. There is no Double St. Nicholas because combining St. Nicholas and Reverse St. Nicholas in the same manner as before will give a method that coincides with either Double Canterbury or Double (Plain) Bob.

6.6   Extents from Groups

Methods and principles facilitate the memorization and actual performance of extents that are based on them by breaking them up into manageable pieces that can be rung on autopilot. Similarly, the actual composition and verification of a b-bell extent are made easier by making use of the natural subdivision of the symmetric group S_b into cosets of a suitable subgroup.

It is interesting to note that 200 years before groups were formally introduced, composers of ringing sequences implicitly used group-theoretic concepts, such as the decomposition of groups into cosets of other groups, to compose and prove their compositions. In fact, the first two books on change ringing, *Tintinnalogia* (1668) [34] and *Campanologia* (1677) [129], contain many examples of ringing sequences, such as extents of the methods Plain Bob Minimus and Plain Bob Doubles and of the principle Stedman

Doubles, whose construction is clearly based on basic group-theoretic insights; see [150] for a more detailed discussion.

Consider the lead ringing array of the extent (=plain course) of Plain Bob Minimus $(AB)^3AC$ in Table 6.1. As elements of the symmetric group S_4, the changes in the first lead (= first column) form a group generated by the two transitions $A = (12)(34)$ and $B = (23)$, which is (isomorphic to) the dihedral group D_4. Hence,

$$D_4 = \{id, A, AB, ABA, (AB)^2, (AB)^2A, (AB)^3, (AB)^3A\}.$$

Here, the group elements (= changes) are listed in the order in which they appear in the ringing sequence. If P is the product $(AB)^3AC = (234)$, with $C = (34)$, then the sets of changes in the second and third leads are the left cosets

$$PD_4 = \{P, PA, PAB, PABA, \dots, P(AB)^3A\},$$
$$P^2D_4 = \{P^2, P^2A, P^2AB, P^2ABA, \dots, P^2(AB)^3A\},$$

of D_4 in the symmetric group.

Again considered as elements of the symmetric group, the changes in the first row of the lead ringing array (= the three changes at the top) form a group that is (isomorphic to) the cyclic group Z_3. Hence,

$$Z_3 = \{id, P, P^2\}.$$

The other rows of the lead ringing array are the right cosets

$$Z_3A, Z_3AB, Z_3ABA, \dots, Z_3(AB)^3A$$

of this subgroup in the symmetric group S_4.

In the following, we translate these observations into procedures for composing extents using left and right cosets of subgroups of symmetric groups. The main references for the material presented in this section are [41], [147], and [149]. See also [49], [92], [93], [148], and [150].

6.6.1 Left Cosets and Plain Bob

We first describe how left cosets can be used to construct extents and have a closer look at extents of Plain Bob, the most popular method on any number of bells, that are constructed in this way.

We start with a subgroup G of order o in S_b. Such a group has $k = b!/o$ left cosets. To construct an extent based on the group, we first try to construct a transition sequence $A_1A_2\cdots A_{o-1}$ whose corresponding sequence of changes consists of exactly the changes in the group G. Considered as a product, the transition sequence is a certain element P' of the symmetric group. In our introductory example $b = 4$, the group is $G = D_4$, its order is $o = 8$, and a working transition sequence is $(AB)^3A$.

Then, we try to construct a sequence of transitions

$$B_1 B_2 \cdots B_k$$

such that

$$A_1 A_2 \cdots A_{o-1} B_1 A_1 A_2 \cdots A_{o-1} B_2 \cdots A_1 A_2 \cdots A_{o-1} B_k$$

is the transition sequence of an extent. A necessary condition for this transition sequence to be an extent is that the k left cosets

$$(P'B_1)(P'B_2) \cdots (P'B_i)G,$$

$i = 1, 2, \ldots, k$, of G in the symmetric group are mutually disjoint and the last of these cosets coincides with the group G. For example, in our introductory example, the extent of Plain Bob Minimus, the second sequence of transitions is CCC.

Plain Bob

Plain Bob is a method that can be defined on any number of bells b greater than 3 and corresponds to the transition sequence

$$(AB)^{b-1} AC,$$

where

$$
\begin{aligned}
A &= (12)(34)\cdots, \\
B &= (23)(45)\cdots, \\
C &= (34)(56)\cdots.
\end{aligned}
$$

The changes in the first lead of its plain course form the elements of a group isomorphic to the dihedral group D_b of order $2b$. Traditionally, any extent of Plain Bob is constructed as outlined above. Here, the first sequence of transitions is always $(AB)^{b-1}A$. The second sequence of transitions is made up of multiple copies of C and of calls that are always made at the ends of leads.

Composers try to use as few different calls as possible. In the case of four bells, no call is required since the plain course of Plain Bob Minimus equals the whole extent. For more bells, composers try to get away as much as possible using the call

$$D = (23)(56)\cdots.$$

As we have seen, in the case of five bells, no further calls are needed. In the case of six or more bells, one more call is necessary; see [144], Section 3. Traditionally, this is the call

$$E = (56)(78)\cdots.$$

In the general setting, once the first transition sequence has been chosen, there is a neat way of expressing finding the second sequence in graph-theoretic terms. We illustrate this in the special case of Plain Bob Doubles (5 bells). In extents based on Plain Bob Doubles, in fact based on any method, the changes at the beginning and the end of a lead both have a 1 in the first position, and no change in the middle of a lead has this property. As we mentioned already, we want to construct the second sequence using the transitions $C = (34)$ and $D = (23)$ only.

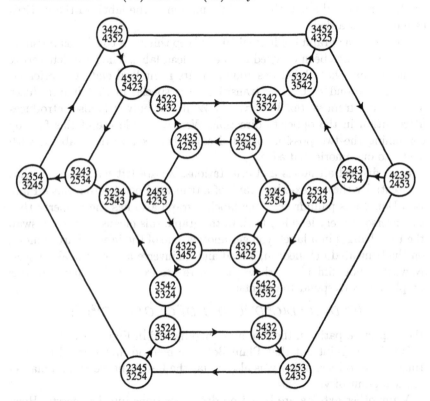

FIGURE 6.10. A graph used for constructing the extents of Plain Bob Doubles. The vertices are labeled with the lead ends and the edges with the plain course transition C (oriented) and the call D (not oriented).

Consider the graph whose vertices are the different possible leads labeled by the two lead ends. Every possible permutation with a leading 1 occurs exactly twice among these labels, once as the front and once as the back lead end. Two vertices V and W are connected by an oriented edge labeled by C or D originating at V, and pointing at W if applying C or D, respectively, to the back lead end in V yields the front lead end in W.

Using this setup, the special extents we are trying to construct are in one-to-one correspondence with the closed oriented paths in this graph (paths that respect the orientation of the oriented edges contained in them), which contain the vertex whose label has rounds as its front lead end and every possible permutation with a leading 1 occurs in exactly one of the labels of the vertices in the path.

Finding an extent in this way is definitely much less complex than finding an extent in the full 5-bell Cayley graph. Just note that the graph has only $4! = 24$ vertices while the full 5-bell Cayley graph has $5! = 120$ vertices; compare Figures 6.8 and 6.9. Furthermore, the path we are looking for has only $(5 − 1)!/2 = 12$ vertices. What is slightly more complicated in the smaller graph is the fact that the information in the labels of the vertices cannot be discarded.

The resulting graph is Figure 6.10. To keep things as concise as possible, the leading 1 has been omitted from every lead label. The lower left vertex is the vertex whose label has rounds as its front lead end. The oriented edges correspond to C. The transition D is special in that it introduces a transition from vertex V to vertex W if and only if it also introduces a transition in the opposite direction. We have made use of this fact by combining the two possible oriented edges between vertices labeled with a D into one unoriented edge.

If we discard all labels and orientations, we are left with a graph that is essentially the vertex-edge graph of a truncated octahedron. Since Plain Bob Doubles is palindromic, the labeled graph also has the property that the inverse of every lead is a lead. In the graph, this means that if you swap the two changes in a label, you get another one of the labels. Furthermore, on the truncated octahedron, a lead and its inverse are antipodal vertices. Now, it is not difficult to show that there are exactly four paths in this graph that correspond to extents:

$$(C^3 D)^3, C^2 D(C^3 D)^2 C, D(C^3 D)^2 C^3, CD(C^3 D)^2 C^2;$$

the respective paths in the graph are highlighted in Figure 6.11.

We should point out that Plain Bob, the method first learned by most ringers due to its simplicity, is also by far the best behaved from a mathematical point of view.

Many other extents are based on decompositions into left cosets. However, the connection is usually not as clear-cut as in the case of extents based on Plain Bob. As we pointed out before, a number of considerations concerning the ringability of a composition are important in practice and may conflict with purely mathematical considerations. Also, it is important to realize that rigorous mathematical considerations only entered the picture after many classics and traditions of change ringing had already been established.

As an example, let us consider the method Grandsire $F(BA)^{b-1}B$, with A and B the same as for Plain Bob and $F = (12)(45)(67) \cdots$, which is defined for any odd number of bells $b \geq 5$. As in the case of Plain Bob, the plain course of Grandsire is based on left cosets of the dihedral group D_b generated again by A and B. However, things are somewhat complicated by the

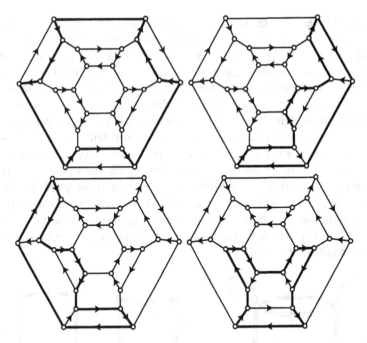

FIGURE 6.11. The four paths corresponding to the four different extents of Plain Bob Doubles.

fact that the method starts off with an F that does not have anything to do with D_b. This has the effect that the first coset consists of change 2 to change $2b + 1$, the second coset consists of change $2b + 2$ to change $4b + 1$, and so on. This also means that, instead of being the first coset as in Plain Bob, the group D_b is the last coset in this decomposition.

Also, the decomposition of the plain course does not quite extend to full coset decompositions of extents of Grandsire, although what remains is still very useful for the effective verification of the extents; see [147], Section 4, for more details. See also [41].

Extents based on the principles Stedman and Erin, again on an odd number of bells b, are based on coset decompositions of S_b into D_3 consisting of six elements. Also, decompositions into the two cosets of the alternating group A_b are apparent in a number of popular extents; see again [41] and [147], Section 4, for more details and examples.

6.6.2 Right Cosets and No-Call Principles

Consider the lead ringing array of the plain course of an arbitrary b-bell method or principle with o working bells. Then, it is clear from the definitions of methods and principles that the changes in the first row of this lead ringing array form a group that is isomorphic to the cyclic group Z_o of order o. If g is the first change of the second lead, then g generates this cyclic group, the first change of the ith lead is g^{i-1}, and all rows of the lead ringing array are right cosets of the cyclic group. If the plain course coincides with an extent—that is, if we are dealing with a *no-call method or principle*—then these right cosets form a partition of the symmetric group S_b. From what we said before, it is clear that all 4-bell methods are no-call methods. For all these methods, the cyclic group in question is Z_3. It is visible as an order 3 symmetry of the Hamiltonian cycles in Figure 6.6.

Right cosets are exactly what we need to construct all no-call b-bell principles. Here is what we have to do. Start by choosing a cyclic subgroup of S_b isomorphic to Z_b (it does not matter which one exactly; see [149], Theorems 2, 3, and 4). The *b-bell Schreier right coset graph* has as its vertices the $(b-1)!$ right cosets of Z_b in S_b. Given any (b-bell) transition A and any right coset R, there is an unoriented edge labeled A that connects R and RA; see the left diagram in Figure 6.12 for a picture of the 4-bell Schreier right coset graph.

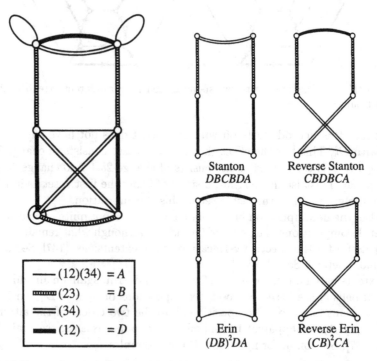

FIGURE 6.12. On the left is the 4-bell Schreier right coset graph. If we assume Z_4 to be generated by the cycle (1243), then Z_4 corresponds to the highlighted vertex in the lower-left corner. On the right are the four Hamiltonian cycles in this graph that correspond to no-call 4-bell principles.

Consider a Hamiltonian cycle in this graph and, starting at the vertex Z_b, record this cycle, as usual, as a sequence of transitions t. Consider this sequence as a product of elements in S_b, and let p be this product. Then, t

is a no-call b-bell principle if and only if p is a cycle of order b (in the group-theoretic sense). Note that every Hamiltonian cycle gives rise to two sequences of transitions corresponding to the two directions in which the cycle can be transversed. If one of these sequences is a no-call principle, then the second sequence is also a no-call principle and the extents corresponding to these sequences are inverses of each other. Furthermore, if t is a no-call principle, then all cyclic permutations of t are also no-call principles. In fact, any no-call b-bell principle arises in this manner from one of the Hamiltonian cycles under consideration.

It is easy to see that there are a total of eight Hamiltonian cycles in the 4-bell Schreier right coset graph depicted in Figure 6.12. We list one of the two different transition sequences corresponding to each of these Hamiltonian cycles.

$$
\begin{aligned}
(DB)^2 DA &= (1342) \\
(CB)^2 CA &= (1342) \\
DBCBDA &= (1243) \\
CBDBCA &= (1243) \\
(DB)^3 &= id \\
(CB)^3 &= id \\
DBCBDB &= (14)(23) \\
CBDBCB &= (14)(23).
\end{aligned}
$$

As you can see, considered as products, exactly four of these are equal to cycles of order 4 (in the group-theoretic sense). These four correspond to the four principles Erin, Reverse Erin, Stanton, and Reverse Stanton that we considered before. The remaining no-call principles on four bells are the "inverses" of these four no-call principles. In fact, up to forming inverses and reverses, there are only two different no-call 4-bell principles, namely Erin and Stanton. Also, the no-call 4-bell principles Erin, Stanton, Reverse Erin, and Reverse Stanton are the only palindromic no-call 4-bell principles.

The minimum number of different transitions needed to build a no-call 5-bell principle turns out to be 3. In [149], Arthur White conducted an in-depth study of the different such "minimal" no-call principles based on the graph-theoretic result outlined above. Also have a look at [147], Section 5, for further ways in which right coset decompositions show up in various popular extents.

6.7   Computers, Bobs, and Singles

Many composers of change ringing sequences are increasingly making use of computers in their work. Using these powerful tools, elements of many distinguished classes of ringing sequences have been completely enumerated and analyzed. For example, on his Web site, the mathematician Alexander Holroyd reports, among other things, that there are exactly 10,792 4-bell extents (162 up to cyclic shifts, reversals, and inversions).

Recently, computers have also been used to solve a century-old problem in change ringing. Change ringers refer to a call as a *bob* or a *single* if it affects an odd or even number of the working bells, respectively. For example, in our discussion of Plain Bob, the call D is a bob and the call E is a single. It is of considerable interest to change ringers to figure out whether, given a method or principle together with a certain set of bobs and singles, an extent of this method or principle can be rung using as calls only the bobs or only the singles. A prominent example of such a problem concerns the 7-bell principle Stedman's Triples,

$$(AB)^2 AC(BA)^2 BC,$$

where $A = (12)(45)(67)$, $B = (23)(45)(67)$, and $C \doteq (12)(34)(56)$. There are 5041 changes in a 7-bell extent. The plain course of this particular principle contains 85 changes. To compose an extent of Stedman Triples, the bob $D = (12)(34)(67)$ and the single $E = (12)(34)$ are used, replacing C in either or both of its occurrences in the principle. Until recently, the most famous unsolved problem in change ringing was the following: Is it possible to ring a full extent of Stedman Triples using only the bob D as a call? In 1994, Colin Wyld [154] finally succeeded in settling this famous problem in the affirmative using a computer; see also [150] page 777.

Not using computers, a number of nonexistence proofs of extents of certain methods and principles have been derived. In 1886, Thompson [138] was able to prove that an extent of Grandsire Triples cannot be rung using only the bob that is traditionally employed for this purpose; see [32] and [41] for a detailed outline of Thompson's proof. Further nonexistence results in the same direction relating to the methods Plain Bob and Grandsire and the principles Stedman and Erin (on any number of bells) have been derived by Rankin [95], [96] and White [147]. For a construction of extents of Plain Bob on b bells, where b is divisible by 4, using only the single $(b - 1, b)$, see [148].

7
Juggling Loose Ends

This chapter is a collection of short articles on juggling- and mathematics-related topics that do not fit into any of the other chapters.

7.1 Does God Juggle?

Recently, it was proved that there may exist a system of three suns of approximately equal masses that move around each other very much like the three balls in a 3-ball cascade. In the following, we give a brief introduction to this kind of "celestial juggling."

The *n-body problem*—that is, the problem of determining the ways in which n punctual masses can move with respect to each other in n-dimensional Euclidean space—is one of the outstanding problems in mathematics. Mathematically, the general n-body problem can be phrased in terms of a system of nonlinear second-order differential equations. Solutions to the n-body problem are suggested by the ways planets, suns, and other celestial bodies move with respect to one another. For example, we know exactly how two bodies can orbit each other. Newton showed that the path followed by each object in such a simple system is a conic section—that is, an ellipse, a parabola, or a hyperbola—one of whose focal points coincides with the center of mass of the system. This includes degenerate solutions in which the two bodies move on a line through the center of mass in which they collide, or from which they get ejected, at some point in time. Of course, these degenerate solutions are just the solutions to the 2-body problem in one dimension. We also know the general solution to the n-body problem in one dimension.

Unlike in the 2-body case, only very few solutions are known that involve dimensions greater than 1 and more than two bodies. In fact, it has been known for a long time that most such systems are chaotic. This means that slight variations in the initial conditions of such a system lead to great changes in its long-term behavior, making any predictions impossible.

The two simplest sets of solutions to the 3-body problem were discovered in 1765 by Euler [39] and in 1772 by Lagrange [69]. Both sets include solutions for all ratios of masses. In all solutions, the bodies move on conic sections. In the case of Euler's solutions, the three bodies are collinear at all times, and the ratios of their distances remain constant. These solutions are unstable and cannot exist in nature. In Lagrange's solutions, the three bodies are the vertices of equilateral triangles at all times. The sun, Jupiter, and one of the so-called Trojan asteroids orbit one another in this manner. As in this real-world example, one of Lagrange's solutions is stable only if one of the bodies is much heavier than the two others.

The most interesting set of solutions from an astronomer's point of view is Hill's set of periodic solutions [15]. In these solutions, two of the bodies are close together, like the Earth and the Moon, and one is far away from these two, just as the Sun is far away from the Earth and the Moon. The first two move on almost circular orbits around their common center of mass, while the third body and this center of mass orbit around the common center of mass of all three bodies, again on almost circular orbits. Hill's solutions also include solutions for all possible ratios of masses.

In the following, we will only be dealing with the special n-body problem in which all bodies are assumed to be of equal mass and move around in a plane. Furthermore, we will focus on periodic solutions to this problem.

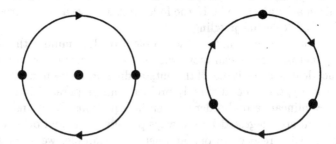

FIGURE 7.1. Two classical solutions to the 3-body problem.

Figure 7.1 shows the simplest among Euler's solutions and the simplest among Lagrange's solutions for the special 3-body problem. In Euler's solution, one of the bodies stays fixed and the two other bodies rotate around it on a circle, with the first body as center, such that at all times the three bodies are on a line. In Lagrange's solution, the three bodies are located at the vertices of an equilateral triangle that rotates around its center.

Recently, Cristopher Moore [85] and Alain Chenciner and Richard Montgomery [23] found a new solution to the special 3-body problem in which the three bodies orbit around each other in a cascade pattern; see Figure 7.2. Just like the ideal 3-ball cascade, this *3-body cascade* is periodic and every one of the bodies follows the same path.

FIGURE 7.2. The 3-body cascade.

Of course, if the new solution is the "3-body cascade," then Lagrange's solution depicted in Figure 7.1 is the *3-body shower* and it is clear how to generalize this solution to an *n-body shower* that is a solution to the special *n*-body problem. In this *n*-body shower, the *n* bodies are located at the vertices of a regular *n*-gon that rotates around its center. Recently, Simó [122] discovered a host of periodic solutions of the special *n*-body problem in which, just as in the 3-body cascade, the bodies travel along the same path in the plane. Simó refers to these new solutions as *choreographies*. Figure 7.3 shows a number of orbits corresponding to these solutions. Note, in particular, that for any odd number of bodies, there is an *n*-body cascade.

It is important to remark that most of the new solutions are known to exist only in universes that are based on (gravitational) potentials of the form

$$f(d) = d^{-b},$$

where $b > 1$, and d stands for "distance." For some, we know that they cannot exist in a Newtonian universe ($b = 1$) such as the one we live in. However, the existence of the 3-body cascade in a Newtonian universe has been proved in [23]. Simó also conjectures the existence of the *n*-body cascade in our universe as well as that of some of the other solutions whose orbits are listed in Figure 7.3.

The 3-body cascade seems to be much more stable than all the other new nontrivial solutions to the special *n*-body problem found so far. Unlike most of these other solutions, minor changes in its initial conditions such as the mass ratios and angular momentum of the three bodies will not lead to solutions in which the three bodies are flying apart. Instead, the new solutions will, in general, stay very close to the original 3-body cascade. If they move away from it, they do so extremely slowly. This implies that there

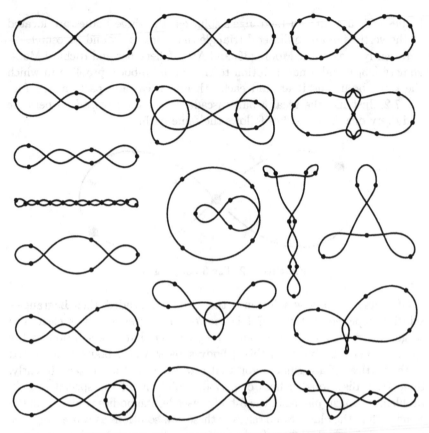

FIGURE 7.3. New solutions to the n-body problem.

may actually be three suns somewhere in our universe that perform a 3-body cascade. Numerical experiments by Douglas Heggie from Edinburgh University place the chances for one such 3-body system to exist in the range of one in one galaxy to one in one universe.

Bodies moving in a plane draw out braids in space-time or, more accurately, in 3-dimensional "plane-time" (two spatial dimensions plus one time dimension = three dimensions). In Section 7.2, we investigate which braids correspond to juggling patterns. Interestingly, the first paper in which the 3-body cascade made its appearance was trying to do the same in the plane n-body problem setting; see [85].

For more detailed mathematical information about the new solutions to the n-body problem, see [22] and [84]. For JAVA animations of some of the new solutions, including the 3-body cascade, see [22]. These JAVA animations are based on *gnuplot* animations by Simó, which can be downloaded from [121]. For some more popular accounts, see the *New Scientist* articles [6] and [87].

7.2 Juggling Braids

Imagine that while you are juggling some juggling pattern in a plane in front of you, you are also jogging backwards at a constant speed[1] and smoke is issuing from your juggling balls. This has the effect that the balls trace their own trajectories in the air just as some airplanes write advertisements in the sky. Figure 7.4 shows what the set of trajectories produced by juggling the 3-ball cascade would look like; see also the cover illustration of this book. In this picture, you, the juggler, move in the direction of the z-axis. The three coordinate axes are visible in the background and intersect in the origin. The positive parts of the z-, y-, and x-axes point to the left, to the upper-right corner, and to the lower-right corner of the picture, respectively. The braid that corresponds to the 3-ball cascade juggled with two hands in the usual manner is the basic 3-stranded braid.

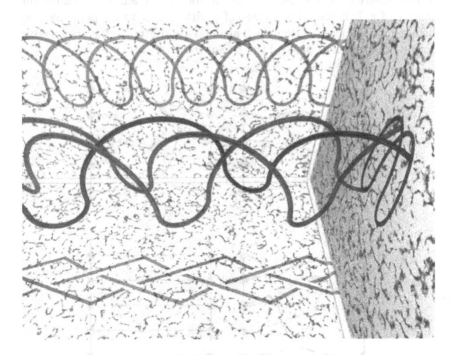

FIGURE 7.4. Juggling the 3-ball cascade while jogging backwards.

We want to show that every finite braid can be juggled. In fact, this is not difficult to show once we explain what exactly we mean by "finite braid" and "juggling a finite braid."

---

[1] *Joggling* is juggling while running. At juggling conventions, various juggling competitions such as 100 meter joggling are held.

Braids and knots play an important role in lower-dimensional topology, the part of mathematics concerned with, among many other things, figuring out the shape of our 3-dimensional universe. Here is a mathematical definition of a finite braid for "pedestrians"; see [133] for precise definitions.

Start with two planes parallel to the xy-plane and call them the left and the right planes. Mark b distinct points on a horizontal line in the left plane and the corresponding b points on the right plane such that corresponding pairs of points have the same xy-coordinates. Add strings that connect the points on the left with the points on the right. This establishes a one-to-one correspondence between the points on the right and the points on the left. However, it is not necessary that corresponding points get connected by a string. We call the resulting configuration a b-*braid* if no two of the strings intersect and every plane parallel to the left and right planes intersects every one of the strings in at most one point; see Figure 7.5 for pictures of four different braids viewed from the top. The braid z^{-1} is called the *inverse* of z. It is the mirror image of z with respect to the plane halfway between the left and right planes. The braid e with straight lines connecting corresponding points is called the *trivial braid*.

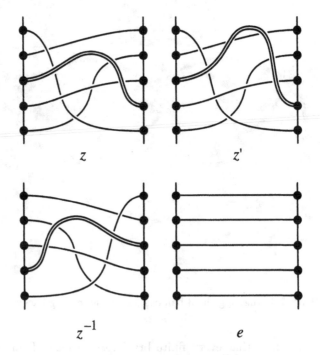

FIGURE 7.5. Four 5-braids.

We consider two braids to be *equivalent*—that is, basically the same—if the strings of the first braid can be deformed in a continuous fashion into the strings of the second braid such that at all times during the transformation process we are dealing with a braid. This means, in particular, that throughout the transformation the end points of the strings stay fixed and at no time during the transformation do strings intersect. The *equivalence class* of a braid consists of all braids equivalent to it. This includes the braid itself.

For example, the braids z and z' are clearly equivalent, as z can be transformed into z' by transforming the doubly drawn string and leaving the rest of the braid fixed. It is possible to *multiply braids*. Figure 7.6 illustrates how this operation is performed in the case of z and z^{-1}.

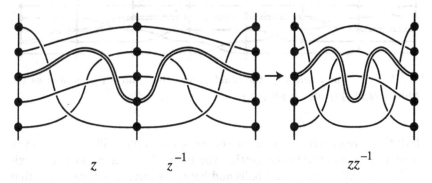

$$z \qquad z^{-1} \qquad zz^{-1}$$

FIGURE 7.6. The product of a braid and its inverse is equivalent to the trivial braid.

What we have to do is to concatenate the two braids and then rescale things horizontally. The diagram also illustrates the fact that if you multiply a braid and its inverse, then the resulting braid is equivalent to the trivial braid. It turns out that the set of equivalence classes of b-braids together with this multiplication is a group, the so-called *braid group* B_b. The equivalence class of the trivial b-braid is the identity element of this group.

Let's number the left end points of the strings $1, 2, \ldots, b$ from top to bottom and the corresponding end points on the right $1', 2', \ldots, b'$. We now define the important *swap braids* $\sigma_{b,i}$, $i = 1, 2, \ldots, b-1$ up to equivalence; see Figure 7.7. In braid $\sigma_{b,i}$, point i is connected with point $(i+1)'$ and point $i+1$ with point i' such that the first string is above the second string. Furthermore, if $j \neq i, i+1$, then j and j' get connected by a straight string. In general, it can be shown that every finite braid is equivalent to one that is a finite product of these swap braids or their inverses. In group-theoretic terms, this just says that the braid group B_b is generated by the $b-1$ swap braids. See [133] for a proof of this important result. Figure 7.7 shows a representation of our sample braid z as the product

$$\sigma_{5,1}\sigma_{5,2}^{-1}\sigma_{5,3}^{-1}\sigma_{5,4}\sigma_{5,3}\sigma_{5,2}\sigma_{5,3}.$$

We can use this result to show very quickly that every finite braid can be juggled. In fact, there are many different settings that we could choose to

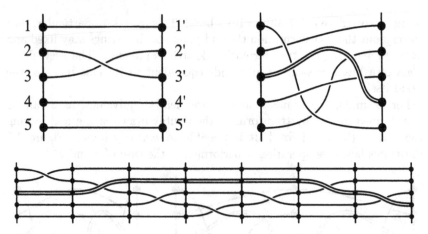

FIGURE 7.7. The braid $\sigma_{5,2}$, the braid z, and the braid z represented as the product $\sigma_{5,1}\sigma_{5,2}^{-1}\sigma_{5,3}^{-1}\sigma_{5,4}\sigma_{5,3}\sigma_{5,2}\sigma_{5,3}$.

show this. We sketch a proof in one of the simplest ones. We leave it to you to adapt it to whatever other setting you prefer. Suppose we want to juggle the braid z. We use as many balls and hands as there are strings in z; that is, five each. Then, we figure out a representation of z in terms of swap braids. We have already found one such representation containing seven factors. We then draw out the graphical counterpart of this representation as in Figure 7.7. This graphical representation is also a set of instructions for the way we have to juggle. We start on the left, all five hands held out above the five end points holding one ball each. Our back is facing towards the right. Now, we walk back seven steps. The length of the steps equals the length of one interval in the representation. At the beginning of every step, every hand throws a ball and at the end every one catches one ball. Hands do not cross. On steps 1 and 2, we perform (simultaneous) throws as in Figure 7.8. Of course, the arrows going up represent self-throws and the parabolic arrows represent throws from one hand to an adjacent one. It should be clear what the remaining five throw combinations are that complete the braid. We could make things even easier by just holding the self-throws.

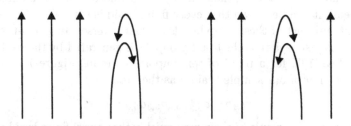

FIGURE 7.8. The first two steps of juggling z.

Here again is our main result of this section.

Juggling Braids

Every finite braid can be juggled.

To conclude this section, here are some more remarks.

First, consider again Figure 7.4 which shows part of the infinite braid that you produce by juggling the 3-ball cascade with two hands while jogging backwards. The orthogonal projections of the braid on the three coordinate planes correspond to different mathematical aspects of juggling. The projection of the braid onto the xy-plane is the slightly deformed infinity sign that we have already encountered. It is the view that you get when you face the juggler, look in the direction of the z-axis, and follow one of the balls around. The projection on the xz-plane looks like the usual projection of a braid onto a plane. This is basically the juggling diagram of the juggling sequence 3 considered as a 2-hand juggling matrix. The projection onto the yz-plane is basically the juggling diagram of the juggling sequence 3 incorporating nonzero dwell time.

As usual, we assume that we have been juggling forever and that we will keep on juggling forever. In this infinite framework, we are also producing infinite braids. However, cutting this braid in two parallel planes that are perpendicular to the direction in which we are moving, we basically get a finite braid as described above. Of course, there are lots of different braids that correspond to the same juggling pattern. However, since we are usually only juggling periodic patterns, there is a finite subset M of such braids such that any other braid produced by cutting the infinite braid as above is equivalent to a product of braids whose factors all belong to M.

There are two more ways in which juggling some kind of pattern produces a braid. First, when you juggle a pattern while standing still, you usually do this such that the balls move around in a vertical plane in front of you. This means that the pattern you are juggling is living in a 2-dimensional universe. In the (2+1)-dimensional space/time universe that extends this 2-dimensional universe, the trajectories of the juggled balls again form a braid; compare this with the remarks at the end of Section 7.1. Of course, this is basically the same braid that we get by jogging backwards. Next (and you can actually do this), attach one end of one string each to the balls that you want to juggle, and attach the loose ends of the strings to the wall. When you then juggle a pattern, you are actually braiding a braid. The resulting braid is equivalent to the one produced by the two other methods of juggling braids. If you actually want to try the last method, make sure that you keep the strings quite long and light and try to juggle a large pattern with exaggerated movements to prevent the strings from getting in the way of juggling.

The physical braid produced by joggling backwards is a perfect record of what we have been doing; that is, even after the juggler is long gone, the trajectories of the balls in the air can be used to reconstruct the pattern that the juggler has been juggling. However, the braid that we get by attaching strings to balls is very malleable, and once you are finished juggling it only keeps a record up to equivalence of braids. It is possible that completely different patterns yield the same braids up to equivalence. As an extreme example, consider two different b-ball simple juggling sequences. Both juggling sequences can be juggled in b columns each; that is, the balls are moving up and down in b columns in front of you. This gives two b-ball patterns that we would consider different. However, the braids produced by these two patterns are both trivial and therefore indistinguishable up to equivalence.

For further thoughts on this topic, see [82].

7.3   Spinning Top of a Palm-Spun Pyramid

You are probably familiar with Chinese exercise balls and how they can be spun in a circle in the palm of a hand by using just the fingers and thumb of this hand; see Figure 7.9. Spinning three balls and a pyramid consisting of four balls in one hand is also possible. Using both hands, a juggler can manipulate up to eight balls in this way, and, by making balls move between hands, cascade patterns with three and five balls can be roll-juggled. These kinds of palm spinning tricks are part of what jugglers refer to as *contact juggling*. As opposed to toss juggling, the objects manipulated in contact juggling usually stay in touch with the juggler's body at all times.

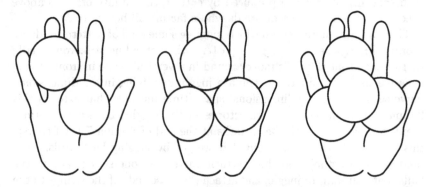

FIGURE 7.9. Spinning two, three, and four crystal balls in the palm of a hand.

Here, we want to focus on spinning the 4-ball pyramid. This very impressive trick is usually performed with four *crystal balls* of equal radius that stay as tightly packed as possible during the spinning movement. Actually, these crystal balls are really unbreakable plastic balls that look, weigh, and cost about as much as real crystal balls of the same size. On closer inspection, you find that, as the base of the pyramid is spinning in the counterclockwise direction, the three balls that form the base of the pyramid each rotate in the clockwise direction, and the ball on top rotates in the counterclockwise direction; see the left-hand diagram in Figure 7.10. Furthermore, by putting some markings on the top ball, we can highlight the fact that it is spinning much faster than the base. Many people find this surprising. In the following, we want to figure out how much faster the top ball spins.

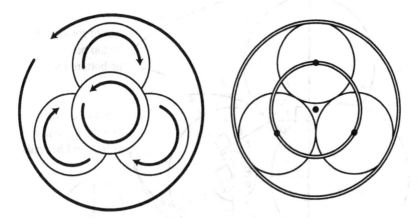

FIGURE 7.10. Spinning a pyramid consisting of four crystal balls.

We may assume that any two of the balls touch at all times and that therefore their four centers are always the vertices of a tetrahedron. Since all the balls at the bottom spin in the same direction, any two of them are moving in opposite directions at the point at which they touch. This means that this trick really only works with balls that have a slippery surface. On the other hand, the top ball and any of the balls at the bottom move in the same direction at the point at which they touch. This means that the trick will still work if we replace the crystal ball on top by a rubber ball of the same size and that there is essentially no slipping happening between the top and any of the bottom balls.

The next thing we observe is that any of the bottom balls rotates around an axis that is tilted away from its vertical symmetry axis. This means that we can replace our palm by a pair of circular rubber rails in which the bottom balls run. Figure 7.11 shows how the rubber rails, one of the bottom balls, and the top ball are linked together (in profile). Then, the value of R in this figure is somewhere between the radii of the two doubly drawn circles in the right-hand diagram in Figure 7.10. We call R the radius of the rubber rail track.

If r denotes the radius of the crystal balls, then simple calculations show that the smaller of the two doubly drawn circles has radius $2r/\sqrt{3}$. Now, it is easy to read off the following relationships between the different angles and distances in Figure 7.11:

$$R = r\left(\tfrac{2}{\sqrt{3}} + \sin\alpha\right), 0 < \alpha < \tfrac{\pi}{2},$$

$$\beta = \arccos\left(\tfrac{1}{\sqrt{3}}\right), \quad \gamma = \pi - (\alpha + \beta),$$

$$b = r\sin\gamma, \qquad t = \tfrac{r}{\sqrt{3}}.$$

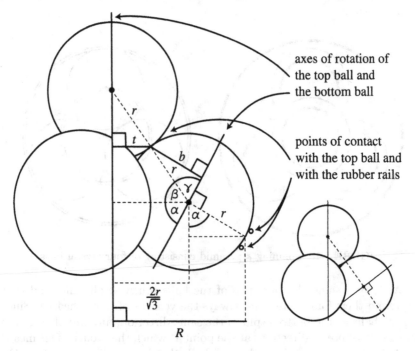

FIGURE 7.11. The linking of the rubber rails, a bottom ball, and the top ball.

The circumferences of the rubber rail track and a ball are $2\pi R$ and $2\pi r$, respectively. Therefore, in the course of one revolution of the base of the pyramid, one of the bottom balls rotates R/r times around its axis. This then translates into $Rb/(rt)$ rotations of the top ball with respect to the pyramid and $Rb/(rt) + 1$ rotations with respect to the hand. We conclude that the top ball spins

$$\frac{Rb}{rt} + 1 = \left(2 + \sqrt{3}\sin\alpha\right)\sin\left(\pi - \arccos\left(\frac{1}{\sqrt{3}}\right) - \alpha\right) + 1$$

times as fast as the bottom of the pyramid. Figure 7.12 shows the graph of this expression considered as a function in the variable α in the allowable range $0 < \alpha < \pi/2$. We see that, at least in theory, the top ball can spin about 4.22 times as fast as the bottom of the pyramid. This corresponds to an angle of roughly $52.6°$.

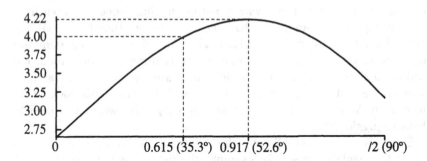

FIGURE 7.12. The top ball spins up to 4.22 times as fast as the base.

When I do this trick, the top ball rotates about four times as fast as the bottom of the pyramid, which roughly corresponds to the special configuration of rotation axes sketched in the lower-right corner of Figure 7.11. In this special case,

$$\alpha = \frac{\pi}{2} - \beta \quad (35.3°).$$

We only note that it is possible to spin two and three rubber balls in one hand, which is much harder than spinning balls that have a slippery surface. The reason is that the balls have to be spun such that no two of them touch at any time (as soon as they touch, they stop moving). It is even possible to spin a pyramid of four rubber balls. Here, no two of the bottom balls touch, and the ball on top actually helps to keep the bottom three from touching. For more information about palm spinning and other contact juggling tricks, see [38].

7.4   Useless Juggling

This chapter is dedicated to fun things to do with juggling numbers. Any one of them can be used to very good effect in lectures on the mathematics of juggling. For the most part, we will focus on simple juggling and leave it to the interested reader to port our results and remarks to other interesting juggling scenarios.

7.4.1   Juggling Words

In this book and most articles on mathematical juggling, numbers such as 4413 are used as abbreviations for juggling sequences. This only works without ambiguity if we agree that we will only be using throws up to height 9. Otherwise, how are we to know whether 20 stands for the sequence 2, 0 or just for the single-element sequence 20?

If we let 'A' stand for a 10-throw, 'B' for an 11-throw, and so forth, we do not have to give up our lazy way of noting juggling sequences, even if we are dealing with throws higher than height 9 (and less than height 36). For example, A4B is a short form of 10, 1, 11. Many of the juggling animators use this convention. Of course, once you start using this convention, you will immediately want to know which words are jugglable. You may have wondered about the somewhat strange way I dedicated this book to my wife, Anu. Well, it turns out that ANU is a jugglable word (you need 21 balls to juggle ANU).

Some people have actually put whole dictionaries through programs that check for jugglability; see, for example, the list of jugglable words in [31] that has been produced this way.

It may not come as too much of a surprise that an APPLE is jugglable. But what about a CAT or a BEEKEEPER (the longest jugglable word I know)? The following sentence consists entirely of jugglable words. Most of these words have something to do with mathematics.

> THUS UNFIT NERDS ABLY BUG A MAD RUSTIC BEEKEEPER PUSH ME TEST, CHECK, PLUS EXTEND MY ABOVE BRIEF COLUMN INDICES PROOF, LEST A PHI PLUS KAPPA LOCUS VORTEXES LINKS MAP WRECK OUR LAST BIG SOLVABLE INNER RANGE ROOT ORDER THEOREM.

Ronald Graham has been reported to ask during his math and juggling lectures (see [24]) whether it is harder to come up with a conjecture for a theorem or to prove it. It so happens that you need 21 balls to juggle THEOREM and 23 to juggle PROOF. So, what is the correct answer to his question?

If your name, just like mine, cannot be juggled like this, you can always juggle it as a one-element multiplex juggling sequence. For example, the word JUGGLE is not jugglable as a simple juggling sequence as above. However, juggling the multiplex juggling sequence [JUGGLE], with letters interpreted as integers as above, is not a problem; see Section 3.5 (look under union). You need $19+30+16+16+21+14=116$ balls to juggle [JUGGLE]. Note that both a word and any of its anagrams are juggled in the same way in this manner.

However, if you prefer simple juggling sequences to multiplex ones, you can still construct a simple juggling sequence that juggles this word. To construct its juggling diagram, label the beat points periodically with the word that you want to juggle as in Figure 7.13. Then, for each of the different letters in the word, draw in one orbit that touches exactly those beat points labeled with the letter associated with the orbit. Remember that this is also one of the ways in which we turned ringing sequences into simple juggling sequences. Of course, lots of different words turn into the same juggling sequence in this way. For example, both the words "juggle" and "nibble" turn into the juggling sequence 661566.

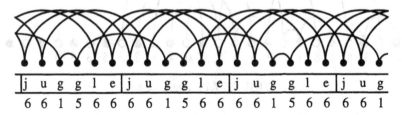

FIGURE 7.13. Any word can be juggled like this.

For further thoughts on this last way of juggling words, see also Wolfgang Schebeczek's article [115] and a paper by Charles Jones [60] in which some links between juggling and telemetry are discussed.

7.4.2 Juggling Rational and Irrational Numbers

In [108], Wolfgang Schebeczek tries to juggle irrational numbers such as e and π. How? Let's consider the decimal expansion of

$$e = 2.718281828459045\ldots.$$

Then we can juggle a 2-throw on beat 1, a 7-throw on beat 2, and so forth. Whenever on a beat no ball is caught but the sequence prescribes a non-zero throw, we pick up a new ball and throw it (we assume that we have an unlimited supply of balls). Otherwise, we stop if the number of balls caught on some beat differs from the number of balls that are supposed to get thrown on this beat. Figure 7.14 illustrates what happens in the case of e. As you can see, nothing goes wrong up to the thirteenth decimal/beat. Then, on the fourteenth beat, a ball lands but the sequence requires that we make a 0-throw—that is, do nothing. Too bad. Still, what is the chance of a random thirteen-digit number being jugglable like this? Not very high, to be sure. In the case of π, things go wrong on the eighth beat. If we "fix" things and change the eighth decimal 5 to a 3, things will only go wrong on the sixteenth beat.

Clearly, every juggling sequence corresponds to a rational number that is jugglable in this way. For example, the juggling sequence 4413 corresponds to 4.41344134413.... Is there an irrational number that is jugglable? The answer is "Yes." Let's construct one such number.

We start with the two 2-ball juggling sequences 2 and 31. Since they are both ground-state, we can concatenate multiple copies of these sequences in whichever way we like to arrive at new juggling sequences. Let 0.100101...

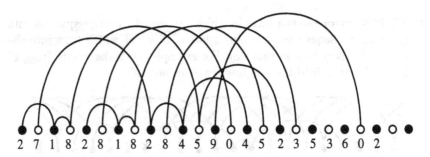

FIGURE 7.14. Juggling e.

be one of the irrational numbers that has only 0s and 1s in its decimal expansion. We replace every 0 in this expansion by a 2 and every 1 by 31. The resulting number 2.312231231 . . . is clearly an irrational number that is jugglable. In fact, using the same trick, it is possible to turn every irrational number into a jugglable irrational number (either use the binary expansion of the irrational number or ten essentially different ground-state juggling sequences to encode the ten digits of the decimal expansion of the number).

7.4.3   Antiballs, Antithrows, and Causal Diagrams

In the last subsection, we convinced ourselves that certain irrational numbers can be juggled. In the following, we describe two different ways in which negative throws can be incorporated into simple juggling sequences in a consistent manner. We call these two different ways *ball/antiball juggling* and *throw/antithrow juggling*. Again, these ideas first popped up on the newsgroup `rec.juggling` [98]. Note that in some articles the two notions are presented as being identical, which is not correct; see, for example, Andrew Conway's hilarious fictitious report on "high energy juggling" [26] in which this mixing of different notions is (probably) intended to add to the humor and to trick the unwary reader.

Ball/Antiball Juggling

To juggle a negative throw in ball/antiball juggling, you need a special ball, called an *antiball*. When thrown, an antiball travels back in time. If an antiball comes in contact with an ordinary ball, the two balls annihilate each other. On the other hand, a ball/antiball pair can materialize out of thin air.

We now consider finite sequences of integers (not necessarily nonnegative) and try to interpret them in the usual manner as sequences of commands for juggling. Figure 7.15 shows the juggling diagram that corresponds to the sequence 25−1−130; that is, 2, 5, −1, −1, 3, 0. Note that, as before, a positive integer i under a beat says that a ball is being thrown to height i and a 0 says that nothing happens. A negative integer i means that an antiball is being *caught* that has been thrown to height $-i$ exactly $-i$ beats in the future. We draw the trajectories of balls right side up and those of antiballs upside down. We call the sequence we started with a ball/antiball juggling sequence if and only if on every beat exactly one of the following happens:

- Nothing happens.

- One ball gets caught and thrown.

- One antiball gets caught and thrown.

- One antiball gets caught, one ball gets caught, and both vanish.

- One antiball and one ball materialize out of thin air and both are thrown.

Check that this is really what happens in Figure 7.15. In particular, we see two beats on which a ball and an antiball collide and vanish. There are also two beats on which a ball/antiball pair materializes.

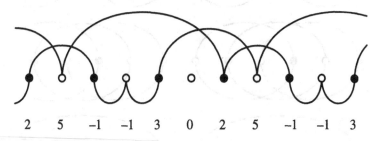

FIGURE 7.15. The juggling diagram of the ball/antiball juggling sequence 25–1–130.

This definition extends our usual notion of a simple juggling sequence in a consistent manner; that is, every simple juggling sequence is a ball/antiball juggling sequence. However, on closer inspection, it turns out that we do not really get anything new. Given any simple juggling sequence, we can replace any of its elements i by $-i$ to arrive at a ball/antiball juggling sequence. For example, in 251130 we can replace the two 1s by -1s to arrive at our example above. In the juggling diagram, this amounts to reflecting the corresponding arcs at the beat line. On the other hand, every ball/antiball juggling sequence clearly arises in this manner.

Throw/Antithrow Juggling

In throw/antithrow juggling, we make a negative throw by throwing an ordinary ball back in time.

Again, we consider finite sequences of integers and interpret them very much like in the case of ball/antiball juggling (juggling diagram and all) except that a negative number i in a sequence means that on the corresponding beat a ball is thrown $-i$ beats into the past.

We call a sequence of integers a throw/antithrow juggling sequence if and only if, as in the case of simple juggling sequences, at most one ball gets caught and thrown on every beat and, if one is caught, it is immediately thrown again. This definition also extends our usual notion of a simple juggling sequence in a consistent manner, and this time we do get something essentially new and interesting. Figure 7.16 shows the juggling diagrams for a number of fun throw/antithrow sequences.

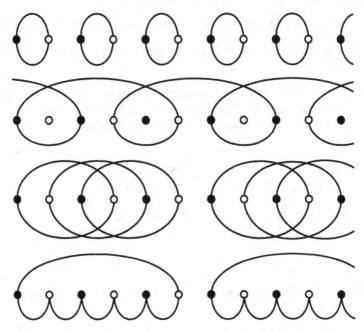

FIGURE 7.16. Juggling diagrams of the four throw/antithrow juggling sequences 1–1, 50–2, 333–3–3–3, and 5–1–1–1–1–1.

You can check for yourself that virtually every result about simple juggling sequences has an extension to a result about throw/antithrow sequences. Here are a few examples:

- *Permutation Test.* A finite sequence of integers s is a throw/antithrow juggling sequence if and only if its associated function ϕ_s is a permutation; see page 22 for a definition of the function ϕ_s. Also, in Section 2.4 we showed that simple juggling sequences correspond to special periodic permutations of the integers. It is easy to see that this correspondence extends naturally to a one-to-one correspondence between the set of all throw/antithrow juggling sequences and the set of all periodic permutations of the integers.

- *Average Theorem.* We are juggling a throw/antithrow juggling sequence. Sometime in between two beats, we count the number of balls in the air such that a ball that travels forwards or backwards in time is counted as 1 ball or −1 ball, respectively. Then, the total number of balls in the air is the average of the elements in the sequence.

- *Site Swaps.* Site swaps turn throw/antithrow juggling sequences into throw/antithrow juggling sequences. More generally, we can apply any of the operations collected in Section 3.5 to throw/antithrow juggling sequences to construct new such sequences.

It is also clear that the *negative* of every throw/antithrow juggling sequence is a throw/antithrow juggling sequence and that the simple juggling sequences are exactly the throw/antithrow juggling sequences all of whose elements are nonnegative.

Causal Diagrams

Throw/antithrow sequences with negative elements may sound like a fairly far-fetched idea to many jugglers. However, the so-called *causal diagrams*, which many jugglers use to describe juggling patterns, have a lot to do with this concept.

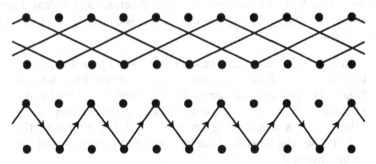

FIGURE 7.17. A juggling diagram and its associated causal diagram.

Figure 7.17 shows the juggling diagram and the so-called *causal diagram* of the 2-hand juggling matrix that is the standard 2-hand interpretation of the simple juggling sequence 3 (what we see here are two representations of the 3-ball cascade). Given a simple b-ball h-hand juggling matrix, $b > h$, its causal diagram is essentially the juggling diagram of the vertical $(-h)$-shift of this juggling matrix. Remember that in order to construct this vertical $(-h)$-shift you just have to subtract h from all entries of the matrix. Of course, if the number of hands is greater than the smallest entry of our matrix, then the vertical $(-h)$-shift will include negative entries, and we will have entered the realm of *throw/antithrow matrices*, which can be defined in analogy to throw/antithrow sequences.[2] To indicate whether an arc in a causal diagram corresponds to a positive or negative number, we provide it with an arrow that points to the right or left, respectively. In our example, the causal diagram is (up to addition of arrows) the juggling diagram of the 2-hand juggling matrix that is the standard 2-hand interpretation of the basic juggling sequence 1.

Figure 7.18 shows the causal diagrams of the 2-hand juggling matrices that are the standard 2-hand interpretations of the juggling sequences 31 and 4413.

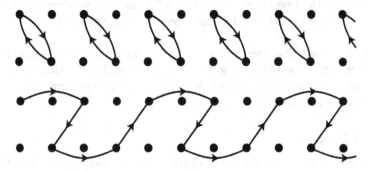

FIGURE 7.18. Causal diagrams associated with the juggling sequences 31 and 4413.

There is a neat way of interpreting causal diagrams as *problem juggling*. Charlie Dancey [29], page 18, describes what is meant by this using the 3-ball cascade.

> At any given moment of a cascade of three objects the juggler has a *problem*—there is one more object in the air than there are hands. The problem approaches the hand and *causes* the juggler to make a throw; so now there is a new problem heading for the other hand. A causal diagram tracks, not the motion of the objects, but the motion of the problem that causes the objects to be thrown.

This interpretation takes some getting used to, but once you start using causal diagrams, you will probably find that very often a causal diagram reflects what you feel when you juggle a particular pattern much better than the juggling diagram. Apart from this, a causal diagram is in general much simpler because, if the juggling diagram consists of b orbits, then the causal diagram consists of only $b - h$ orbits. For example, in Figure 7.17, the juggling diagram consists of three orbits, whereas the causal diagram consists of only one orbit. Of course, the intuitive idea of the "number

---

[2] *Vertical shifts.* Convince yourself, using the permutation test, that vertical shifts of simple juggling matrices are really throw/antithrow matrices.

of orbits" can also be extended in a straightforward way to account for degenerate cases like the first causal diagram in Figure 7.18.

Club jugglers in particular find that causal diagrams are the best way of recording passing patterns. This is partly because of the reasons we just mentioned but also because often things can be arranged in such a way that throws that are thrown with single spins, double spins, triple spins, and so forth end up being represented by 1-beat, 2-beat, and 3-beat arrows, respectively. Originally, it was in this context that causal diagrams were invented and used by Martin Frost; see [42] and [43].

For a very nice introduction to causal diagrams, including many good examples, see [29], page 18. Note that actual drawings of causal diagrams of passing patterns in the literature usually look a lot simpler than juggling diagrams, as they also incorporate a number of shortcuts based on common features of passing patterns.

7.5 Juggling and Math Stories

The following is a collection of stories that will be of interest to most mathematically minded jugglers.

7.5.1 Riddle

In [8] and [118], different versions of an old riddle are recounted. Here is the version included in [8].

> To complete a delivery of munitions, a 148-pound man has to traverse a high, creaking bridge that can support only 150 pounds. He is carrying three one-pound cannonballs and has time for only one trip across. How can he cross the bridge and make his delivery in time?

The solution that is usually given is that while the man is crossing the bridge, he is juggling a 3-cannonball cascade. However, juggling the cannonballs will not work because, although the juggler is only holding a maximum of two cannonballs in his hands at any time during the crossing, the average weight of the juggler and the cannonballs would still be 151 pounds, and this weight has to be supported via the juggler's feet by the bridge. This means that, unless the bridge is really short and he is able to throw the balls all the way across the bridge, there will be a moment during the crossing when the weight limit of the bridge is exceeded.

On the other hand, the following bowling solution may be more realistic. As the man approaches the bridge, he bowls the cannonballs in rapid succession ahead of himself across the bridge, follows them across, and gathers them up on the other side, all the while not breaking his stride in order not to lose any precious time. Of course, even that may not work since just the force of a running 148-pound man may exceed at some point the weight limit of the bridge and make it collapse. We only mention in passing that there are jugglers capable of juggling three full-sized bowling balls for close to one hundred throws.

7.5.2 Lord Valentine's Castle

The following passage from the fantasy novel "Lord Valentine's Castle" by Robert Silverberg [120] has been cited a number of times in the juggling literature; see, for example, [118]. It certainly summarizes nicely the feelings of many jugglers.

It is the story of Valentine, a man who is trying to reconstruct his forgotten past. He teams up with a band of jugglers and finds, among other things, that he is a juggling genius. The band includes the two human jugglers Sleet and Carabella and some alien four-armed jugglers.

> Carabella returned, bearing a great many colored rubber balls that she juggled briskly as she crossed the yard. When she reached Valentine and Sleet she flipped one of the balls to Valentine and three to Sleet, without breaking stride. Three she retained.
>
> 'Not knives?' Valentine asked.
>
> 'Knives are showy things. Today we deal in fundamentals,' Sleet said. 'We deal in the philosophy of the art. Knives would be a distraction.'
>
> 'Philosophy?'
>
> 'Do you think juggling's a mere trick?' the little man asked, sounding wounded. 'An amusement for the gapers? A means of picking up a crown or two at a provincial carnival? It is all those things, yes, but first it is a way of life, a friend, a creed, a species of worship.'
>
> 'And a kind of poetry,' said Carabella.
>
> Sleet nodded, 'Yes, that too. And a mathematics. It teaches calmness, control, balance, a sense of placement of things and the underlying structure of motion. There is a silent music to it. Above all there is discipline. Do I sound pretentious?'
>
> 'He means to sound pretentious,' Carabella said. There was mischief in her eyes. 'But everything he says is true. Are you ready to begin?'
>
> Valentine nodded.

Other novels in which juggling plays an important role are "Haroun and the Seas of Stories" by Salmon Rushdie [100], "The Physician" by Noah Gordon [47], and "The Clown of God: An Old Story" by Tomie De Paola [30].

7.5.3 Famous Juggler-Mathematicians

Juggling appeals to many mathematically minded people and many, perhaps even the majority of, serious amateur jugglers are people with a mathematical background—mathematicians, computer experts, majors in physics, and the like. A number of famous mathematicians are also expert jugglers. The most famous among these is undoubtedly Ronald Graham, past president of both the American Mathematical Society (1993) and the International Juggler's Association (1972). Ronald Graham is one of the most influential combinatorists alive, whose work has earned him numerous distinctions and prizes. He is former chief scientist for AT&T Laboratories, member of the American Academy of Sciences, and professor of mathematics at the University of California—San Diego. He has also contributed significantly to mathematical juggling by coauthoring a number of fundamental papers in this line of research and giving numerous math-of-juggling talks; see [4] and [116] for excellent accounts of Graham's multifaceted life. See also [24], [56], and [160].

In the previous chapters, we already encountered Claude Shannon's name a number of times. Claude Shannon (1916–2001) was one of the most influential mathematicians and computer scientists of the twentieth century. Only through his discoveries did digital communication as we know it today become possible. Among jugglers, he is best known as the man who built the first juggling robot and proved the first juggling-related theorems. At Bell Labs, he is remembered for riding a unicycle down its corridors and juggling all the while. In fact, he was also one of the founding members of the Unicycling Society of America. For more information about Claude Shannon, see Sections 4.6 and 5.2 as well as [79], [80], [117], and [118].

7.6 Further Reading

In the following, we mention some topics that complement those covered in this book and give references for the interested reader.

If you are interested in the mathematics of control and dynamical systems that is used to model juggling and build robots that perform various juggling and juggling-related tasks such as unicycling, balancing, and so forth, start by checking out the references given in Subsection 5.2.3. If you are interested in a revolutionary new method of balancing multiple linked objects, one on top of the other, make sure you read [1], [2], and [3].

Also have a look at some of the research done by movement scientists such as Peter Beek; see [8] for a popular account as well as [7], [9], [10], [91], and the references given there. From a juggling-math perspective, one of the most interesting topics investigated by movement scientists is the modeling of the different ways in which humans and animals perform periodic tasks that involve a number of different limbs. For example, using coupled nonlinear oscillators, they are able to explain why jugglers prefer to perform 2-hand juggling either by throwing simultaneously with both hands or having the hands take turns throwing such that one hand is throwing at the same time as the other hand is catching. See also the article [132] by Ian Stewart and [25] for other good introductions to this fascinating topic (juggling is not mentioned though).

Finally, for a general list of references relating to "circus science," see [109].

Appendix
Stereograms of Hamiltonian Cycles

In Subsection 6.5.1, we showed that a b-bell extent corresponds to an oriented Hamiltonian cycle in the b-bell Cayley graph. For $b = 4$, the most symmetric representation of this graph is related to the truncated octahedron. In this representation, the vertices of the graph are the vertices of the truncated octahedron. Its edges are the edges of this Archimedean solid plus all diagonals of the square faces.

In this appendix, we exhibit stereograms of the graph and Hamiltonian cycles in this graph that correspond to the extents of methods and principles that we considered in Subsection 6.5.2.

Did you ever succeed in viewing a *Magic Eye* random dot stereogram? If so, you have already mastered either the *parallel technique* or the *cross-eyed technique*. Either of these techniques can also be used to view the following stereograms. The two images in a stereogram are the two views of a spatial object as seen by your left and right eyes. What you have to figure out is a way to merge these two flat pictures into a spatial picture. For this, one of your eyes has to focus on one of the pictures while the other eye has to focus on the second picture.

With the parallel technique, your right eye has to focus on the right picture and your left eye has to focus on the left picture. This is achieved by staring "through" the picture plane at a point at infinity; see the left diagram in Figure A.1.

With the cross-eyed technique, your left eye has to focus on the right picture and your right eye has to focus on the left picture. To achieve this, hold a pencil between your eyes and the picture plane. Focus on the tip of the pencil with both your eyes. While keeping your eyes fixed on its tip, slowly move the pencil towards the picture plane. Once the correct position is reached, the spatial image should come into view; see the right-hand diagram in Figure A.1. Here, the pencil is positioned where the two arrows cross.

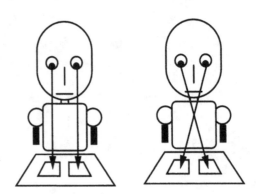

FIGURE A.1. Two techniques for viewing stereograms.

If you have mastered both techniques, you will notice that the three-dimensional pictures get turned "inside out" when you switch from one technique to the other. If you are nearsighted and prefer the parallel technique, try viewing the stereograms without your glasses on.

Although most people have no (serious) problems viewing stereograms with either the parallel or the cross-eyed technique, some people, especially those with eye problems other than near- or farsightedness, find it impossible to make stereograms work. If you experience problems, a mechanical stereo viewer may be the solution. These are available from specialized dealers such as stereoscopy.com. About a century ago, stereograms were very much in fashion. Therefore, it is also worth asking around in your local antique shops whether they have any stereogram viewers from that period.

Figure A.2 is a stereogram of the 4-bell Cayley graph modeled onto the truncated octahedron. The edges that are thicker than the rest are the ones labeled $C = (34)$. Among the "thin" edges, the ones that are diagonals of square faces, edges of square faces, and edges of only hexagon faces are the edges labeled $A = (12)(34)$, $D = (12)$, and $B = (23)$, respectively. The vertex that the arrow points at is the common starting point of the Hamiltonian cycles corresponding to the various extents under consideration.

A partition of the Cayley graph into squares is visible in Figure A.2. Partitions into octagons and hexagons are displayed in Figures A.3 and A.4. Remember that the Bob extents are based on the octagons (that is, the "great circles" of the truncated octahedron); see Figures A.5, A.6, and A.7. Partitions into hexagons are apparent in the Court and Erin extents; see Figures A.11, A.12, A.13, A.16, and A.17. A partition into squares is apparent in Figure A.10 (Double Canterbury).

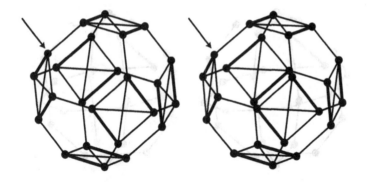

FIGURE A.2. The 4-bell Cayley graph.

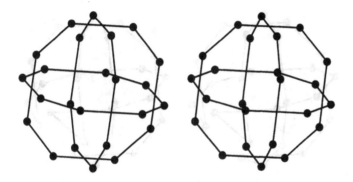

FIGURE A.3. Partition into octagons.

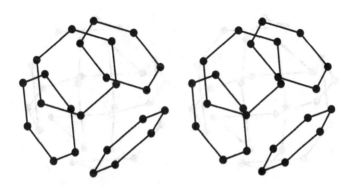

FIGURE A.4. Partition into hexagons.

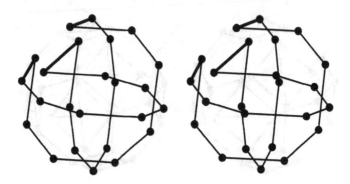

FIGURE A.5. Plain Bob $((AB)^3 AC)^3$.

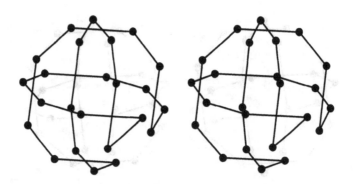

FIGURE A.6. Reverse Bob $(ABAD(AB)^2)^3$.

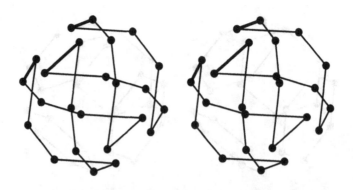

FIGURE A.7. Double Bob $(ABADABAC)^3$.

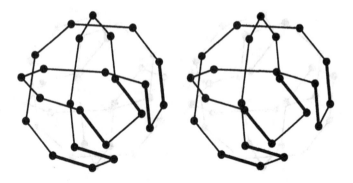

FIGURE A.8. Canterbury $(ABCDCBAB)^3$.

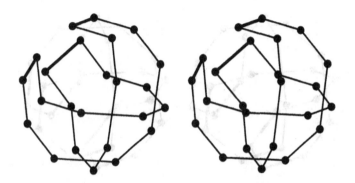

FIGURE A.9. Reverse Canterbury $(DB(AB)^2DC)^3$.

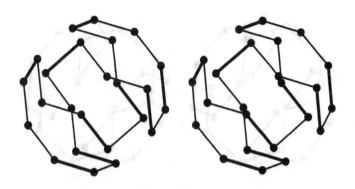

FIGURE A.10. Double Canterbury $(DBCDCBDC)^3$.

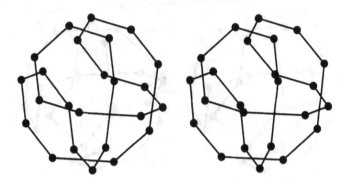

FIGURE A.11. Single Court $(DB(AB)^2DB)^3$.

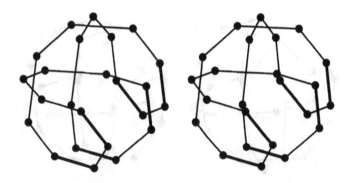

FIGURE A.12. Reverse Court $(AB(CB)^2AB)^3$.

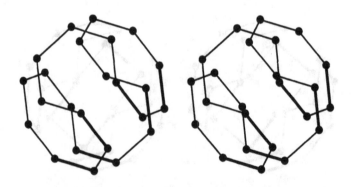

FIGURE A.13. Double Court $(DB(CB)^2DB)^3$.

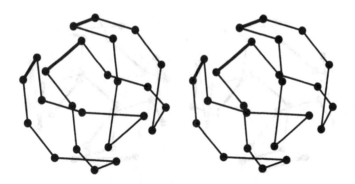

FIGURE A.14. St. Nicholas $(DBADABDC)^3$.

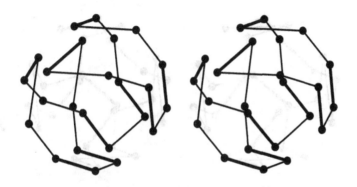

FIGURE A.15. Reverse St. Nicholas $(ABCDCBAC)^3$.

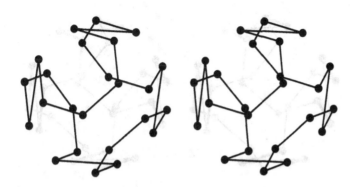

FIGURE A.16. Erin $((DB)^2DA)^4$.

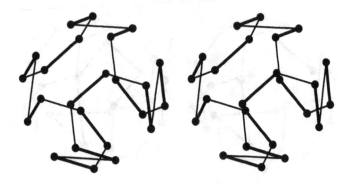

FIGURE A.17. Reverse Erin $((CB)^2CA)^4$.

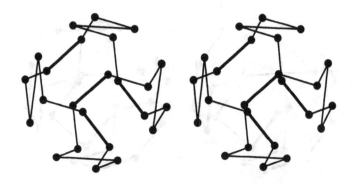

FIGURE A.18. Stanton $(DBCBDA)^4$.

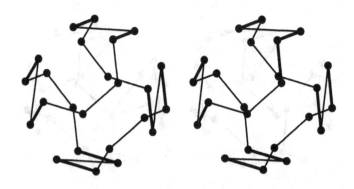

FIGURE A.19. Reverse Stanton $(CBDBCA)^4$.

References

[1] Acheson, David. A pendulum theorem. *Proc. R. Soc. Lond. A* 443 (1993), 239–245.

[2] Acheson, David; Mullin, Tom. Upside-down pendulum. *Nature* 366 (18 November 1993), 215–216.

[3] Acheson, David; Mullin, Tom. Ropy magic. *New Scientist* (21 February 1998), 32–33.

[4] Albers, Donald J. A nice genius. *Math Horizons* (November 1996), 18–23.

[5] Alvarez, Francisco. Juggling—Its History and Greatest Performers. Online book, 1984 (available online at the Juggling Information Service [62] in the directory **/books**).

[6] Appel, David. Celestial swingers. *New Scientist* (4 August 2001), 36–39.

[7] Beek, Peter. Juggling Dynamics. Free University Press, Amsterdam, 1989.

[8] Beek, Peter; Lewbel, Arthur. The science of juggling. *Scientific American* (November 1995), 92–97 (available online at the Juggling Information Service [62] in the directory **/papers**).

[9] Beek, Peter; Turvey, Michael. Temporal patterning in cascade juggling. *J. Exp. Psychol.* 18 (1992), 934–937.

[10] Beek, Peter; Van Santvoord, Anthony. Learning the cascade juggle: A dynamical systems analysis. *Journal of Motor Behavior* 24 (1992), 85–94.

[11] Beever, Ben. Siteswap Ben's Guide to Juggling Patterns. Published by the author at 93 Delph Lane, Delph, Oldham, Lancashire, OL3 5UP, U.K., 2000.

[12] Berggren, J. Lennart. Episodes in the Mathematics of Medieval Islam. Springer-Verlag, New York, 1986.

[13] Boyce, Jack. Article on complements of juggling sequences on the newsgroup rec.juggling (19 May 1993). Accessible via the Juggling Information Service [62].

[14] Boyce, Jack. Article on prime juggling sequences on the newsgroup rec.juggling (4 June 1998). Accessible via the Juggling Information Service [62].

[15] Brown, Ernest W. An introductory treatise on the lunar theory, Sections II.18–24 and Chapter XI, Cambridge University Press, Cambridge, U.K., 1896.

[16] Buhler, Joe; Graham, Ronald. Fountains, showers, and cascades. *The Sciences* (January–February 1984), 44–51.

[17] Buhler, Joe; Graham, Ronald. A note on the binomial drop polynomial of a poset. *J. Combin. Theory Ser. A* 66 (1994), 321–326.

[18] Buhler, Joe; Graham, Ronald. Juggling drops and descents. Proceedings of the Organic Mathematics Workshop, 12–14 December 1995 (available online at the Juggling Information Service [62] in the directory /papers).

[19] Buhler, Joe; Eisenbud, David; Graham, Ronald; Wright, Colin. Juggling drops and descents. *Amer. Math. Monthly* 101 (1994), 507–519.

[20] Carlitz, Leonard. *q*-Bernoulli numbers and polynomials. *Duke Math. Journal* 15 (1948), 987–1000.

[21] Carstens, Ed. Mathematics of juggling. Online publication, 1992 (available online at the Juggling Information Service [62] in the directory /papers).

[22] Casselman, Bill. A new solution to the *n*-body problem—and more. Online publication of the American Mathematical Society, 2001 (available at www.ams.org/new-in-math/cover/orbits1.html).

[23] Chenciner, Alain; Montgomery, Richard. A remarkable periodic solution of the three-body problem in the case of equal masses. *Ann. of Math.* 152 (2000), 881–901.

[24] Cole, K.C. Plenty of balls in the air. *Los Angeles Times* (20 May 1999), Home Edition, Section: Metro, p. B2.

[25] Collins, J.J.; Stewart, I.N. Coupled nonlinear oscillators and the symmetries of animal gaits. *J. Nonlinear Sci.* 3 (1993), 349–392.

[26] Conway, Andrew. Article on "high energy juggling" on the newsgroup `rec.juggling` (3 May 1994). Accessible via the Juggling Information Service [62].

[27] The Council's Decisions. The Central Council of Church Bell Ringers, Morphet, 1989.

[28] Dancey, Charlie. Charlie Dancey's Encyclopedia of Ball Juggling. Butterfingers, Bath, 1994.

[29] Dancey, Charlie. Charlie Dancey's Compendium of Club Juggling. Butterfingers, Bath, 1995.

[30] De Paola, Tomie. The Clown of God: An Old Story. Harcourt Brace, New York, 1978.

[31] De Rooij, Steven. Article on jugglable words on the newsgroup `rec.juggling` (13 September 1998). Accessible via the Juggling Information Service [62].

[32] Dickinson, D.J. On Fletcher's paper "Campanological groups." *Amer. Math. Monthly* 64 (1957), 331–332.

[33] Donahue, Bill. Jugglers now juggle numbers to compute new tricks for ancient art. *New York Times* (16 April 1996), p. C1.

[34] Duckworth, R. Tintinnalogia. W. Godbid, London, 1668.

[35] Dworkin, Morris. Factorization of the cover polynomial. *J. Combin. Theory Ser. B* 71 (1997), 17–53.

[36] Ehrenborg, Richard; Readdy, Margaret. Juggling and applications to q-analogues. *Discrete Math.* 157 (1996), 107–125.

[37] Engels, André; Mauw, Sjouke. Why men (and octopuses) cannot juggle a four ball cascade. *J. Recreational Math.*, to appear.

[38] Ernest, James. Contact Juggling. Second edition. Ernest Graphics Press, Seattle, 1991.

[39] Euler, Leonard. De motu rectilineo trium corporum se mutuo attrahentium, Novi Commentarii Academiae Scientarum Petropolitanae 11 (1767), pp. 144–151, in: Oeuvres, Seria Secunda tome XXV Commentationes Astronomicae (page 286).

[40] Finnigan, Dave. The Complete Juggler. Jugglebug, 1992.

[41] Fletcher, T.J. Campanological groups. *Amer. Math. Monthly* 63 (1956), 619–626.

[42] Frost, Martin. More site swaps for passing. *Juggler's World* 45 no. 2 (Summer 1994), 34–36.

[43] Frost, Martin. Passing lots o'club. *Juggler's World* 45 no. 4 (Winter 1994–95), 32–35.

[44] Giduz, Bill. The joy of zero-g juggling. *Juggler's World* 37 no. 2 (Summer 1985), 4–6.

[45] Gillen, Billy. Remember the force Hassan! *Juggler's World* 38 no. 2 (Summer 1986), 9–10 (available online at the Juggling Information Service [62] in the directory /jw/86/2).

[46] Goodwin, Graham; Graebe, Stefan; Salgado, Mario. Control System Design. Prentice-Hall, Upper Saddle River, New Jersey, 2001. See also the accompanying Web pages that include nice implementations of various control solutions of the inverted pendulum problem at murray.newcastle.edu.au/control.

[47] Gordon, Noah. The Physician. Macmillan, London, 1988.

[48] Gregorio, M.; Ahmadi, M.; Buehler, M. Experiments with an electrically actuated planar hopping robot. In: Experimental Robotics III. Edited by Yoshikawa, T. and Mayazaki, F., Springer, New York, 1994.

[49] Griffiths, Deryn. Twin bob plan compositions of Stedman triples: Partitioning of graphs into Hamiltonian subgraphs as used in bell-ringing. *Bull. Inst. Combin. Appl.* 16 (1996), 65–76.

[50] Hall, Marshall. A combinatorial problem on abelian groups. *Proc. Amer. Math. Soc.* 3 (1952), 584–587.

[51] Hall, Marshall. Combinatorial Theory. John Wiley & Sons, New York, 1967.

[52] Harrison, John. The Tower Handbook. The Central Council of Church Bell Ringers, Morphet, 1997.

[53] Hickerson, Dean. Article on reordering finite sequences of integers into juggling sequences on the newsgroup `rec.juggling` (12 January 2001). Accessible via the Juggling Information Service [62].

[54] Hickerson, Dean. Article on reordering finite sequences of integers into juggling sequences on the newsgroup `rec.juggling` (9 February 2001). Accessible via the Juggling Information Service [62].

[55] Hickerson, Dean. Article on reordering finite sequences of integers into juggling sequences on the newsgroup `rec.juggling` (6 March 2001). Accessible via the Juggling Information Service [62].

[56] Hoffman, Paul. The man who loves only numbers. *Atlantic Monthly* (November 1987).

[57] Holden, C. Juggling robot. *Science* 251 (15 February 1991), 742.

[58] Horgan, John. Ronald L. Graham: Juggling act. *Scientific American* 276 (March 1997), 21–22.

[59] Johnson, Dave. The veteran neophyte: Paper juggling. *Juggler's World* 47 no. 1 (Spring 1995), 43–45.

[60] Jones, Charles H. Telemetry and juggling. Proceedings of the International Telemetering Conference vol. XXXVI (2000), paper 00-20-2.

[61] JOTA (Jongliertheoretische Arbeitsgemeinschaft). Juggling theory. Part 1: The DAB-Theorum and other serious illnesses. *Kaskade* 29 (March 1993), 35–37.

[62] Juggling Information Service (JIS), one of the main sources of juggling-related resources on the net at `www.juggling.org`.

[63] Kalvan, Jack. The optimal juggler. Online publication, 1996 (available online at the Juggling Information Service [62] in the directory `/papers`).

[64] Kalvan, Jack. The human limits: How many objects can be juggled. Online publication, 1997 (available online at the Juggling Information Service [62] in the directory `/papers`).

[65] Kemper, Steve. The magical motion of Michael Moschen. *Smithsonian* 26 no. 5 (August 1995), 38–47 (an abstract of this article is available online at the Juggling Information Service [62] in the directory `/papers`).

[66] Kincanon, Eric. Juggling and the theorist. *Physics Teacher* 28 (April 1990), 221–223.

[67] Knutson, Allen. Siteswap FAQ. Version 2.0. Online publication, 10 November 1993 (available online at the Juggling Information Service [62] in the directory /help/siteswap).

[68] Knutson, Allen; Levine, Matt; Warrington, Greg. Juggling by the numbers. This online publication is available at www.cs.unc.edu/~helser/juggler-0.81/clean/main.html.

[69] Lagrange, Joseph-Louis. Essai sur le Problème des Trois Corps. Prix de l'Académie Royale des Sciences de Paris, tome IX. In: Vol. 6 of Oeuvres (page 292).

[70] Lewbel, Arthur. Memorable tricks and a numbers formula (The Academic Juggler). *Juggler's World* 39 no. 4 (Winter 1987/88), 31 (available online at the Juggling Information Service [62] in the directory /jw/87/4).

[71] Lewbel, Arthur. A short lesson in thought & gravity (The Academic Juggler). *Juggler's World* 40 no. 2 (Summer 1988), 31.

[72] Lewbel, Arthur. Getting back around your own clubs (The Academic Juggler). *Juggler's World* 40 no. 4 (Winter 1988/89), 34.

[73] Lewbel, Arthur. The Academic Juggler. *Juggler's World* 41 no. 2 (Summer 1989), 27.

[74] Lewbel, Arthur. Clubs have come a long way (The Academic Juggler). *Juggler's World* 41 no. 4 (Winter 1989/90), 32.

[75] Lewbel, Arthur. Freefalling, or jugglers' many commonalities (The Academic Juggler). *Juggler's World* 42 no. 3 (Fall 1990), 25 (available online at the Juggling Information Service [62] in the directory /jw/90/3).

[76] Lewbel, Arthur. Picture that (The Academic Juggler). *Juggler's World* 43 no. 3 (Fall 1991), 12–13.

[77] Lewbel, Arthur. A free offer, a call to teachers, and the invention of juggling notations (The Academic Juggler). *Juggler's World* 45 no. 4, (Winter 1993/94), 34–35.

[78] Lewbel, Arthur. The Early History of Juggling. *Juggler's World* 47 no. 4 (Winter 1995/96), 36–37 (available online at the Juggling Information Service [62] in the directory /papers under the title "Research in juggling history").

[79] Lewbel, Arthur. Obituary for Claude Elwood Shannon. *Juggle* (May/June 2001), 17.

[80] Liversidge, Anthony. Profile of Claude Shannon. In: Claude Elwood Shannon. Collected papers. Edited by Sloane, N. and Wyner, A. IEEE Press, New York, 1993, pp. xix–xxxii (originally *Omni Magazine* (August 1987)).

[81] MacDonald, Alan; Schoenberg, Ben; Schebeczek, Wolfgang. World records. *Kaskade* 49 (1998), 23.

[82] Margalit, Dan; Picciotto, Neil "Fred"; Llobrera, Joseph; Babineau, Sarah. Topology and juggling (available at www.brown.edu/Students/OHJC/topology/).

[83] Miyamoto, H.; Gandolfo, F.; Gomi, H.; Schaal, S.; Koike, Y.; Rieka, O.; Nakano, E.; Wada, Y.; Kawato, M. A Kendama learning robot based on a dynamic optimization principle. In: Proceedings of the International Conference On Neural Information Processing 2, Hong Kong, 24–27 September 1996, pp. 938–942.

[84] Montgomery, Richard. A new solution to the three-body problem. *Notices Amer. Math. Soc.* 48 (2001), 471–481.

[85] Moore, Cristopher. Braids in classical gravity. *Phys. Rev. Lett.* 70 (1993), 3675–3679.

[86] Moreton, Wilfried F. Method Construction. The Central Council of Church Bell Ringers, Morphet, 1996.

[87] Muir, Hazel. Take your partners for the planet waltz. *New Scientist* (14 April 2001), 16.

[88] Oudshoorn, Willem. Article giving a proof that the throws in a simple juggling sequence are determined by the states it visits on the newsgroup rec.juggling (18 July 1998). Accessible via the Juggling Information Service [62].

[89] Oudshoorn, Willem. Article giving an example of two juggling sequences associated with the same set of juggling states on the newsgroup rec.juggling (20 July 1998). Accessible via the Juggling Information Service [62].

[90] Peterson, Ivars. Juggling by design. Online column "Ivars Peterson's MathTrek" (2 August 1999) of the Mathematical Association of America (available at www.maa.org).

[91] Post, A.; Daffershofer, A.; Beek, P. Principal components in three-ball cascade juggling. *Biological Cybernetics* 82 (2000), 143–152.

[92] Price, Brian D. Mathematical groups in campanology. *Math. Gaz.* 53 (1969), 129–133.

[93] Price, Brian D. The Composition of Peals in Parts. Published by the author, 19 Snarsgate Street, London W10 6QP, U.K., 1994.

[94] Probert, Martin. Four Ball Juggling. Published by Veronika Probert, U.K., 1995.

[95] Rankin, R.A. A campanological problem in group theory. *Proc. Cambridge Philos. Soc.* 44 (1948), 17–25.

[96] Rankin, R.A. A campanological problem in group theory. II. *Proc. Cambridge Philos. Soc.* 62 (1966), 11–18.

[97] Rapaport, Elvira S. Cayley color groups and Hamilton lines. *Scripta Math.* 24 (1959), 51–58.

[98] rec.juggling. The archive of this main newsgroup for jugglers is accessible via the Juggling Information Service [62].

[99] Roaf, Dermot; White, Arthur. Ringing the changes: Bells and mathematics. preprint.

[100] Rushdie, S. Haroun and the Seas of Stories. Granta Books (in association with Viking Penguin), London, 1990.

[101] Sabin, Katie; Schoenberg, Ben. Numbers juggling—the state of the art. *Juggle* (January/February 2000), 16–25.

[102] Sachkov, Vladimir N. Combinatorial Methods in Discrete Mathematics. Encyclopedia of Mathematics and its Applications 55, Cambridge University Press, Cambridge, 1996.

[103] Sanderson, J. (ed.) Change Ringing: The History of an Ancient Art. Volume 1. The Central Council of Church Bell Ringers, Morphet, 1987.

[104] Sanderson, J. (ed.) Change Ringing: The History of an Ancient Art. Volume 2. The Central Council of Church Bell Ringers, Morphet, 1992.

[105] Sanderson, J. (ed.) Change Ringing: The History of an Ancient Art. Volume 3. The Central Council of Church Bell Ringers, Morphet, 1994.

[106] Sayers, Dorothy L. The Nine Tailors. Victor Gollancz, London, 1934.

[107] Schaal, Stefan; Atkeson, Christopher. Open loop stable control strategies for robot juggling. In: Proceedings of the IEEE International Conference on Robotics and Automation, Volume 3, Atlanta, Georgia, 1993, IEEE Press, New York, pp. 913–918.

[108] Schebeczek, Wolfgang. Verbesserung der Zahl π. *Kaskade* 39 (Autumn 1995), 29 (with some corrections under the title "π for beginners" in *Kaskade* 40 (1995), 27).

[109] Schebeczek, Wolfgang. Physics goes to the circus. *Kaskade* 42 (Summer 1996), 14–16.

[110] Schebeczek, Wolfgang. Look, mum, no rider. *Kaskade* 42 (Summer 1996), 17.

[111] Schebeczek, Wolfgang. Multi-hand cascades: No silly thing. Article on the newsgroup rec.juggling (1 July 1998). Accessible via the Juggling Information Service [62].

[112] Schebeczek, Wolfgang. Fuofua kau moua. The juggling girls of the Pacific islands. Part 1. *Kaskade* 51 (Autumn 1998), 12–15.

[113] Schebeczek, Wolfgang. Kita'ita'i ake te rama. The juggling girls of the Pacific islands. Part 2. *Kaskade* 52 (Winter 1998), 14–17.

[114] Schebeczek, Wolfgang. Tiria mai taku pei. The juggling girls of the Pacific islands. Part 3. *Kaskade* 53 (Spring 1999), 32–35.

[115] Schebeczek, Wolfgang. The robots are coming. *Kaskade* 62 (Summer 2001), 32–35.

[116] Schecter, Bruce. Ronald L. Graham. In: Mathematical people. Profiles and interviews. Edited by Albers, D. and Alexanderson, G. Birkhäuser, Boston, 1985, pp. 109–117.

[117] Shannon, Claude E. Claude Shannon's no-drop juggling diorama. In: Claude Elwood Shannon. Collected papers. Edited by Sloane, N. and Wyner, A. IEEE Press, New York, 1993, pp. 847–849.

[118] Shannon, Claude E. Scientific aspects of juggling. In: Claude Elwood Shannon. Collected papers. Edited by Sloane, N. and Wyner, A. IEEE Press, New York, 1993, pp. 850–864.

[119] Sheng, Zaiquan; Yamafuji, Kazuo. Realization of a human riding a unicycle by a robot. Proceedings of the 1995 IEEE International Conference on Robotics and Automation, Volume 2, IEEE Press, New York, 1995, pp. 1319–1326.

[120] Silverberg, R. Lord Valentine's Castle. Voyager (in association with Harper Collins), 1999.

[121] Simó, Carles. A series of 42 animations of n-body choreographies which run under the UNIX program *gnuplot* (available online at www.maia.ub.es/dsg/nbody.html).

[122] Simó, Carles. New families of solutions in n-body problems. Proceedings of the ECM 2000 (Barcelona, July 10–14), to appear.

[123] Simpson, Charlie. Juggling on paper. *Juggler's World* 37 no. 4 (Winter 1986), 31.

[124] Sloane, Neil J.A. Online Encyclopedia of Integer Sequences. Sequence A047996 gives the number of necklaces. Available at www.research.att.com/~njas/sequences/index.html.

[125] Sommers, B. John. Juggling as performing mathematics. *Co-Evolution Quarterly* (Summer 1980), 125–131 (available online at the Juggling Information Service [62] in the directory /papers).

[126] Stadler, Jonathan D. Article on magic juggling sequences on the newsgroup rec.juggling (26 August 1994). Accessible via the Juggling Information Service [62].

[127] Stadler, Jonathan D. Schur Functions, Juggling, and Statistics on Shuffled Permutations. Ph.D. thesis, Ohio State University, Columbus, Ohio, 1997.

[128] Stadler, Jonathan D. Stanley's shuffling theorem revisited. *J. Combin. Theory A* 88 (1999), 176–187.

[129] Stedman, Fabian. Campanalogia. W. Godbid, London, 1677.

[130] Steingrimmson, Einar. Permutation Statistics of Indexed and Poset Permutations. Ph.D. thesis, MIT, Cambridge, Massachusetts, 1991.

[131] Stewart, Ian. Juggling by numbers. *New Scientist* (18 March 1995), 34–38.

[132] Stewart, Ian. Mathematical recreations: Why Tarzan and Jane can walk in step with the animals that roam the jungle. *Scientific American* (April 1991), 88–91.

[133] Stillwell, John C. Classical Topology and Combinatorial Group Theory. Graduate Texts in Mathematics 72, Springer-Verlag, New York and Berlin, 1980.

[134] Storer, Dave. A written notation for describing ball juggling tricks. *International Juggler's Association Newsletter* 30 (March/April 1978), 7.

[135] Strong, Todd. Siteswaps (Teach in), *Juggle* (March/April 2001), 35–39.

[136] Tiemann, Bruce; Magnusson, Bengt. The physics of juggling. *Physics Teacher* 27 (1989), 584–589.

[137] Tiemann, Bruce; Magnusson, Bengt. A notation for juggling tricks. A LOT of juggling tricks. *Juggler's World* 43 no. 2 (Summer 1991), 31–33 (available online at the Juggling Information Service [62] in the directory /jw/91/2 under the title 'Can you please write that pattern down').

[138] Thompson, W.H. A note on Grandsire Triples. London, 1886. (Reprinted in: Snowdon, W. Grandsire, London, 1905, a revision of: Snowdon, J. Grandsire, London, 1888.)

[139] Tritz, Gerry. Soft-spoken techno-fan applies math to juggling through computer program. *Juggler's World* 46 no. 1 (Spring 1994), 24–25.

[140] Truzzi, Marcello. Notes toward a history of juggling. *Bandwagon* 18 no. 2 (March–April 1974) (available online at the Juggling Information Service [62] in the directory /papers).

[141] Voigt, Joachim. Juggling on paper. *Kaskade* 25 (March 1992), 16–17.

[142] Waldmann, Johannes. Article about retrieving juggling throws from juggling states on the newsgroup rec.juggling (20 July 1998). Accessible via the Juggling Information Service [62].

[143] Walker, Jeff. Variations for numbers jugglers. *Juggler's World* 34 no. 1 (January 1982), 11.

[144] White, Arthur T. Ringing the changes. *Math. Proc. Cambridge Philos. Soc.* 94 (1983), 203–215.

[145] White, Arthur T. Graphs, Groups and Surfaces. Second edition. North-Holland Mathematics Studies 8. North-Holland Publishing Co., Amsterdam, 1984.

[146] White, Arthur T. Ringing the changes. II. Tenth British combinatorial conference (Glasgow, 1985). *Ars Combin.* 20 A (1985), 65–75.

[147] White, Arthur T. Ringing the cosets. *Amer. Math. Monthly* 94 (1987), 721–746.

[148] White, Arthur T. A Hamiltonian construction in change ringing. Eleventh British Combinatorial Conference (London, 1987). *Ars Combin.* 25 B (1988), 257–264.

[149] White, Arthur T. Ringing the cosets. II. *Math. Proc. Cambridge Philos. Soc.* 105 (1989), 53–65.

[150] White, Arthur T. Fabian Stedman: The first group theorist? *Amer. Math. Monthly* 103 (1996), 771–778.

[151] White, Arthur T.; Wilson, Robin. The hunting group. *Math. Gaz.* 79 (1995), 5–16.

[152] Wright, Colin. SiteSwaps. How to write down a juggling pattern: A guide for the perplexed. Online publication, 1996 (available online at the Juggling Information Service [62] in the directory /help/siteswap).

[153] Wright, Colin. Article on who invented site swaps on the newsgroup rec.juggling (25 May 1995). Accessible via the Juggling Information Service [62].

[154] Wyld, Colin. Around and about. *The Ringing World* (27 January 1995), 104.

[155] Yam, Yeung; Song, Jingyan. Extending Shannon's theorem to a general juggling pattern. *Stud. Appl. Math.* 100 (1998), 53–66.

[156] Zetie, Ken. Juggling notation for non-geeks. Part 1. *Kaskade* 37 (Spring 1995), 16–18.

[157] Zetie, Ken. Juggling notation for non-geeks. Part 2. *Kaskade* 38 (Summer 1995), 30–31.

[158] Ziethen, Karl-Heinz; Allen, Andrew. Juggling, the Art and its Artists. Werner Rausch & and Werner Lüft Inc., Berlin, 1985.

[159] Ziethen, Karl-Heinz. Die Kunst der Jonglerie. Henschelverlag, Berlin, 1988.

[160] Zimmer, Carl. Circus science (acrobats, jugglers and contortionists). *Discover* 17 (February 1996), 56–63.

Index

0, 118
01234, 36
1, 118
2, 118
31, 119
312, 119
330, 119
[33]33, 122
[333], 122
40, 119
423, 120
441, 119, 121
4413, 121
501, 120
50505, 121
51, 121
5111, 119
52512, 121
53, 122
531, 121
534, 123
552, 123
55500, 121
5551, 123
60, 121

600, 119
61111, 119
633, 123
71, 123
711111, 119
7131, 120
73131, 121
7333, 122
8000, 119
9313131, 121
933333, 122

Abu Sahl al-Kuhi, 3
accuracy, 130
algorithm
    flattening, 20
array
    auxiliary, 31
    ringing, *see* ringing array
average test, 16
average theorem
    converse of, 29
    for multihand juggling, 89
    for multiplex juggling, 67
    for simple juggling, 15

average theorem (continued)
for throw/antithrow juggling,
195

ball
bell, 146
crystal, 186
labeled, 112
Beek, Peter, 200
bells
ring of, 141
working, 154
Beni Hassan, 2
bob, 176
Boppo, see Tiemann, Bruce
braid, 182
Buhler, Joe, 38

call, 157, 176
campanology, 141
Canterbury, 149, 156, 205
card
juggling, see juggling card
Carstens, Ed, 5, 6, 85
cascade, 11, 107
3-body, 179
next-hand, 107
previous-hand, 107
reverse, 13
Chalcraft, Adam, 5
change, 141, 147
change ringing, 141
Chenciner, Alain, 179
Cinquevalli, Paul, 4
club, 4, 129
coefficient
Gaussian, 73
concatenation, 82
conductor, 157
Conway, Andrew, 5, 192
cosets
left, 169
right, 173
Cuchulainn, 2
cycle, 160

Hamiltonian, 160, 201

Dancey, Charlie, 123, 196
Day, Mike, 5
Decisions of the Central Council
of Church Bell Ringers,
158
diagram
causal, 195
juggling, see juggling diagram
domino pattern, 108
Double Bob, 149, 156, 204
Double Canterbury, 149, 156, 205
Double Court, 149, 156, 206
duality, 97, 108

Ehrenborg, Richard, 38, 73
Eisenbud, David, 38
elevation of ringing sequence, 152
equivalent, 182
Erin, 154, 149, 156, 207
Euler, Leonard, 178
extension of state graph, 58
extent, 143
of principle or method, 157

flash, 121, 132
Flying Karamazov Brothers, 112
fountain, 11, 107
reverse, 13
same-hand, 107
frequency, 102
total, 105
Frost, Martin, 197
function
Euler, 52, 109
juggling, see juggling function
Möbius, 40

Gatto, Anthony, 4, 132
Graham, Ronald, 38, 138, 190
Grandsire Doubles, 155
Grandsire Triples, 176
graph
Cayley, 159, 171, 201

Schreier right coset, 174
site swap, 21
state, *see* state graph
ground state, 47
group
    affine Weyl, 42
    alternating, 173
    braid, 183
    cyclic, 169
    dihedral, 169
    symmetric, 151

half-shower, 122
Hall, Marshall, 30
Heggie, Douglas, 180
height
    of juggling function, 15
    of throw, 9
Hickerson, Dean, 5, 30
hiko, 3
history of juggling, 2
Hoffmann, Martin 6
Holroyd, Alexander, 176
h-throw, 9
hunt, 155
    plain, 155

inverse
    of juggling sequence, 25, 82
    of ringing sequence, 143, 152
inversion, 41

joculare, 3
JoePass!, 6, 88, 135
joggling, 181
Jongl, 6
Juggle, 5
JuggleAnim, 6, 51
JuggleKrazy, 6
JuggleMaster, 6
JugglePro, 5, 6, 85
Juggler's World, 137
juggling
    balls and antiball, 192
    balls and hands, 110

billiard ball, 124
bounce, 126
braids, 181
contact, 186
history of, 2
multihand, 85
multiplex, 65
practical, 117
problem, 196
robot, 127, 199
simple, 7
throw and antithrow, 193
uniform, 96
words, 189
zero-gravity, 124
juggling cards
    for multihand juggling, 92
    for multiplex juggling, 68
    for simple juggling, 38
juggling diagram, 9
    contracted, 93
juggling function, 14
Juggling Information Service, 5
juggling matrix, 85
    b-ball h-hand of height k, 90
    cyclic, 95
    distributed, 94
    maximal prime, 91
    operations for transforming a, 92
    simple, 94
juggling pattern, *see also* pattern
    baby, 121
    basic, 11
    multiplex, 8, 65
    simple, 7
    uniform, 96, 101
juggling sequence, 4
    b-ball h-hand, 111
    complement of, 83
    concatenation, 82
    contracted, 93
    excited state, 47
    ground-state, 47
    inverse of, 25, 82

juggling sequence (continued)
    magic, 35
    minimal, 9
    multiplex, 65
    number of multiplex, 68
    number of simple, 40
    operations for transforming a,
        81, 92
    prime, 50
        maximal, 52, 77
        trivial, 52, 78
    scramblable, 34
    simple, 7
    union of several, 83
    weight of, 42, 73
juggling state
    b-ball of height h, 44
    generalized, 112
    multihand, 90
    multiplex, 75
    simple, 44

Kara, Michael, 4
Klimak, Paul, 5
Knutson, Allen, 5, 20, 121

Lagrange, Joseph-Louis, 178
Lan Zi, 2
lead, 154
Levine, Matt, 6
Lewbel, Arthur, 5, 126, 136
lining up of props, 132
Lipson, Andrew, 6
loop
    maximal prime, 52, 77, 91
        length of, 57, 80, 91
    prime, 50, 77
    trivial prime, 52, 78

Magic Eye, 201
[MA][GNU]S, 6
Magnusson, Bengt, 5, 129, 135,
    136
matrix, see also juggling matrix
    transition 62

Matsuoka, Ken, 6
method, 144, 154
    no-call, 174
Method Writer, 146
Montgomery, Richard, 179
Moore, Cristopher, 179
multihand notation, 5, 85

necklace
    multiplex, 77
    number of multiplex, 78
    number of simple, 52
    simple, 51
Newton, Isaac, 177
Nine Tailers, The, 145
number
    Fibonacci, 159
    Galois, 73
    intertwining, 74
    Stirling, 74

open loop control strategy, 128
otedama, 3
Otteson, Steve, 5
Oudshoorn, Willem Rein, 5, 49

palindromic, 158
partition, 114
    maximal, 114
    number of partitions of a set,
        74
    of Cayley graph, 163, 166, 202
    of state graph into necklaces,
        51, 77
    of symmetric group into cosets,
        174
passing, 85
pattern, see also juggling pattern
    baby juggling, 121
    cascade, see cascade
    domino, 108
    flash, see flash
    fountain, see fountain
    multiplex, 8, 65
    passing, see passing

pistons, *see* pistons
shower, *see* shower
simple, 8
simultaneous, 95
snake, *see* snake
peal, 145
pendulum
    inverted, 128
period, 9
permutation test
    for multihand juggling, 90,
    for multiplex juggling, 67
    for simple juggling, 22
    for throw/antithrow juggling,
        194
pistons, 12
Plain Bob, 143, 149, 170, 204
plain course, 154
principle, 144, 154
    no-call, 174
Probert, Martin, 5, 28, 48, 122,
    149
problem
    3-body, 178
    n-body, 177
Procedure, Pick a Pattern, 28

qualify, 132

Rabbi Shimon ben Gamaliel, 2
Rappo, Karl, 3
Rastelli, Enrico, 4
Readdy, Margaret, 38, 73
rec.juggling, 4, 30, 34, 36, 49, 192
records
    juggling, 132
    ringing, 145
reflection
    simple, 43
Reverse Bob, 149, 156, 204
Reverse Canterbury, 149, 156, 205
Reverse Court, 149, 156, 206
Reverse Erin, 149, 156, 208
reverse of ringing sequence, 152
Reverse Stanton, 149, 156, 208

Reverse St. Nicholas, 149, 156, 207
Riebesel, Werner, 6
ring of bells, 141
ringing array, 143, 152
    lead, 154
ringing sequence, 142
    elevation of, 152
    reverse of, 152
    vertical shift of, *see* shift
robot
    blind, 128
    seeing, 128
Rooij, Steven, 5
rounds, 142

Sayers, Dorothy, 145
scaling, 82
Schebeczek, Wolfgang, 5, 107, 191,
    191
schedule
    landing, 44
self-dual, 97
sequence, *see also* juggling sequence
    b- of period *p*, 20
    juggling, *see* juggling sequence
    qualifying, 30
    ringing, *see* ringing sequence
    transition, 151
series
    Poincaré, 43
Shannon, Claude, 5, 96, 126, 127,
    199
Shannon's theorems, 96, 103
shift
    cyclic
        of juggling sequence, 19, 81
        of ringing sequence, 152
    vertical
        of juggling sequence, 23, 83
        of ringing sequence, 152
        of state graph, 58
shower, 3, 13
    3-body, 179
    half-, 122
side step, 93

Simó, Carles, 180
single, 176
Single Court, 149, 156, 206
site swap, 4, 19, 82, 195
snake, 121
Soest, Theo van, 146
solution
    Euler's, 178
    Hill's, 178
    Lagrange's, 178
spin, 132
Stadler, Jon, 5, 36
Stanton, 149, 156, 208
state
    excited, 47
    ground, 29, 47
    juggling, see juggling state
state graph
    complement of, 59
    extension of, 58
    multihand, 90
    multiplex, 75
    simple, 44
states determine throws result
    for multihand juggling, 90
    for multiplex juggling, 81
    for simple juggling, 49
Stedman Doubles, 157, 169
Stedman Triples, 176
stereogram, 201
Stewart, Ian, 200
St. Nicholas, 149, 156, 207
subgraph, 51
superball, 126

swinging
    club, 4
synchronicity property, 96

technique
    cross-eyed, 201
    parallel, 201
tenor, 141
test
    average, see average test
    permutation, see permutation
        test
test vector, 22
Thompson, 176
throw, 9
Tiemann, Bruce, 5, 129, 135, 136
time
    dwell, 96, 130
    flight, 96
    total dwell, 104
    total flight, 104
    total vacant, 104
    vacant, 97
Tonga, 3
treble, 141

Waldmann, Johannes, 5, 49
Walker, Jeff, 5
Warrington, Greg, 6
weight, 42, 73
Westerboer, Wolfgang, 6
White, Arthur, 141, 175
Wimsey, Lord Peter, 145
Wright, Colin, 5, 6, 38, 138